INTRODUCTORY NON-EUCLIDEAN GEOMETRY

Henry Parker Manning

DOVER PUBLICATIONS, INC.
Mineola, New York

Bibliographical Note

This Dover edition, first published in 1963 and reprinted in 2005, is an unabridged republication of the work originally published in 1901 by Ginn and Company, Boston, under the title *Non-Euclidean Geometry.*

International Standard Book Number: 0-486-44262-4

Manufactured in the United States of America
Dover Publications, Inc., 31 East 2nd Street, Mineola, N.Y. 11501

PREFACE

Non-Euclidean Geometry is now recognized as an important branch of Mathematics. Those who teach Geometry should have some knowledge of this subject, and all who are interested in Mathematics will find much to stimulate them and much for them to enjoy in the novel results and views that it presents.

This book is an attempt to give a simple and direct account of the Non-Euclidean Geometry, and one which presupposes but little knowledge of Mathematics. The first three chapters assume a knowledge of only Plane and Solid Geometry and Trigonometry, and the entire book can be read by one who has taken the mathematical courses commonly given in our colleges.

No special claim to originality can be made for what is published here. The propositions have long been established, and in various ways. Some of the proofs may be new, but others, as already given by writers on this subject, could not be improved. These have come to me chiefly through the translations of Professor George Bruce Halsted of the University of Texas.

I am particularly indebted to my friend, Arnold B. Chace, Sc.D., of Valley Falls, R. I., with whom I have studied and discussed the subject.

HENRY P. MANNING.

Providence, January, 1901.

iii

CONTENTS

INTRODUCTORY
NON-EUCLIDEAN
GEOMETRY

NON–EUCLIDEAN GEOMETRY

INTRODUCTION

THE axioms of Geometry were formerly regarded as laws of thought which an intelligent mind could neither deny nor investigate. Not only were the axioms to which we have been accustomed found to agree with our experience, but it was believed that we could not reason on the supposition that any of them are not true. It has been shown, however, that it is possible to take a set of axioms, wholly or in part contradicting those of Euclid, and build up a Geometry as consistent as his.

We shall give the two most important Non-Euclidean Geometries.* In these the axioms and definitions are taken as in Euclid, with the exception of those relating to parallel lines. Omitting the axiom on parallels,† we are led to three hypotheses; one of these establishes the Geometry of Euclid, while each of the other two gives us a series of propositions both interesting and useful. Indeed, as long as we can examine but a limited portion of the universe, it is not possible to prove that the system of Euclid is true, rather than one of the two Non-Euclidean Geometries which we are about to describe.

We shall adopt an arrangement which enables us to prove first the propositions common to the three Geometries, then to produce a series of propositions and the trigonometrical formulæ for each of the two Geometries which differ from

* See Historical Note, p. 93.   † See p. 91.

that of Euclid, and by analytical methods to derive some of their most striking properties.

We do not propose to investigate directly the foundations of Geometry, nor even to point out all of the assumptions which have been made, consciously or unconsciously, in this study. Leaving undisturbed that which these Geometries have in common, we are free to fix our attention upon their differences. By a concrete exposition it may be possible to learn more of the nature of Geometry than from abstract theory alone.

Thus we shall employ most of the terms of Geometry without repeating the definitions given in our text-books, and assume that the figures defined by these terms exist. In particular we assume :

I. *The existence of straight lines determined by any two points, and that the shortest path between two points is a straight line.*

II. *The existence of planes determined by any three points not in a straight line, and that a straight line joining any two points of a plane lies wholly in the plane.*

III. *That geometrical figures can be moved about without changing their shape or size.*

IV. *That a point moving along a line from one position to another passes through every point of the line between, and that a geometrical magnitude, for example, an angle, or the length of a portion of a line, varying from one value to another, passes through all intermediate values.*

In some of the propositions the proof will be omitted or only the method of proof suggested, where the details can be supplied from our common text-books.

CHAPTER I

PANGEOMETRY

I. PROPOSITIONS DEPENDING ONLY ON THE PRINCIPLE OF SUPERPOSITION

1. Theorem. *If one straight line meets another, the sum of the adjacent angles formed is equal to two right angles.*

2. Theorem. *If two straight lines intersect, the vertical angles are equal.*

3. Theorem. *Two triangles are equal if they have a side and two adjacent angles, or two sides and the included angle, of one equal, respectively, to the corresponding parts of the other.*

4. Theorem. *In an isosceles triangle the angles opposite the equal sides are equal.*

Bisect the angle at the vertex and use (3).

5. Theorem. *The perpendiculars erected at the middle points of the sides of a triangle meet in a point if two of them meet, and this point is the centre of a circle that can be drawn through the three vertices of the triangle.*

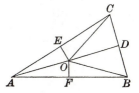

Proof. Suppose EO and FO meet at O. The triangles AFO and BFO are equal by (3). Also, AEO and CEO are equal.

Hence, *CO* and *BO* are equal, being each equal to *AO*. The triangle *BCO* is, therefore, isosceles, and *OD* if drawn bisecting the angle *BOC* will be perpendicular to *BC* at its middle point.

6. Theorem. *In a circle the radius bisecting an angle at the centre is perpendicular to the chord which subtends the angle and bisects this chord.*

7. Theorem. *Angles at the centre of a circle are proportional to the intercepted arcs and may be measured by them.*

8. Theorem. *From any point without a line a perpendicular to the line can be drawn.*

Proof. Let *P'* be the position which *P* would take if the plane were revolved about *AB* into coincidence with itself. The straight line *PP'* is then perpendicular to *AB*.

9. Theorem. *If oblique lines drawn from a point in a perpendicular to a line cut off equal distances from the foot of the perpendicular, they are equal and make equal angles with the line and with the perpendicular.*

10. Theorem. *If two lines cut a third at the same angle,*

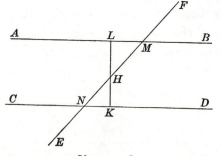

that is, so that corresponding angles are equal, a line can be drawn that is perpendicular to both.

Proof. Let the angles *FMB* and *MND* be equal, and through *H*, the middle point of *MN*, draw *LK* perpendicular to *CD*; then *LK* will also be perpendicular to *AB*. For the two triangles *LMH* and *KNH* are equal by (3).

11. Theorem. *If two equal lines in a plane are erected perpendicular to a given line, the line joining their extremities makes equal angles with them and is bisected at right angles by a third perpendicular erected midway between them.*

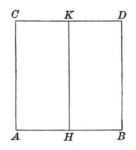

Let *AC* and *BD* be perpendicular to *AB*, and suppose *AC* and *BD* equal. The angles at *C* and *D* made with a line joining these two points are equal, and the perpendicular *HK* erected at the middle point of *AB* is perpendicular to *CD* at its middle point.

Proved by superposition.

12. Theorem. *Given as in the last proposition two perpendiculars and a third perpendicular erected midway between them; any line cutting this third perpendicular at right angles, if it cuts the first two at all, will cut off equal lengths on them and make equal angles with them.*

Proved by superposition.

Corollary. *The last two propositions hold true if the angles at A and B are equal acute or equal obtuse angles, HK being* *perpendicular to AB at its middle point. If AC = BD, the angles at C and D are equal, and HK is perpendicular to CD at its middle point ; or, if CD is perpendicular to HK*

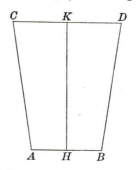

at any point, K, and intersects AC and BD, it will cut off equal distances on these two lines and make equal angles with them.

II. PROPOSITIONS WHICH ARE TRUE FOR RESTRICTED FIGURES

The following propositions are true at least for figures whose lines do not exceed a certain length. That is, if there is any exception, it is in a case where we cannot apply the theorem or some step of the proof on account of the length of some of the lines. For convenience we shall use the word *restricted* in this sense and say that a theorem is true for restricted figures or in any restricted portion of the plane.

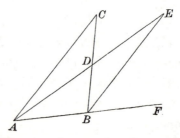

1. Theorem. *The exterior angle of a triangle is greater than either opposite interior angle* (Euclid, I, 16).

Proof. Draw AD from A to the middle point of the opposite side and produce it to E, making $DE = AD$. The two triangles ADC and EBD are equal, and the angle FBD, being greater than the angle EBD, is greater than C.

Corollary. *At least two angles of a triangle are acute.*

2. Theorem. *If two angles of a triangle are equal, the opposite sides are equal and the triangle is isosceles.*

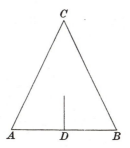

Proof. The perpendicular erected at the middle point of the base divides the triangle into two figures which may be made to coincide and are equal. This perpendicular, therefore, passes through the vertex, and the two sides opposite the equal angles of the triangle are equal.

3. Theorem. *In a triangle with unequal angles the side opposite the greater of two angles is greater than the side opposite the smaller; and conversely, if the sides of a triangle are unequal the opposite angles are unequal, and the greater angle lies opposite the greater side.*

4. Theorem. *If two triangles have two sides of one equal, respectively, to two sides of the other, but the included angle of the first greater than the included angle of the second, the third side of the first is greater than the third side of the second; and conversely, if two triangles have two sides of* one equal, respectively, to two sides of the other, but the third side of the first greater than the third side of the second, the angle opposite the third side of the first is greater than the angle opposite the third side of the second.

5. Theorem. *The sum of two lines drawn from any point to the extremities of a straight line is greater than the sum of two lines similarly drawn but included by them.*

6. Theorem. *Through any point one perpendicular only can be drawn to a straight line.*

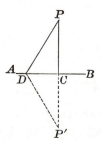

Proof. Let P' be the position which P would take if the plane were revolved about AB into coincidence with itself. If we could have two perpendiculars, PC and PD, from P to AB, then CP' and DP' would be continuations of these lines and we should have two different straight lines joining P and P', which is impossible.

Corollary. *Two right triangles are equal when the hypothenuse and an acute angle of one are equal, respectively, to the hypothenuse and an acute angle of the other.*

7. Theorem. *The perpendicular is the shortest line that can be drawn from a point to a straight line.*

Corollary. *In a right triangle the hypothenuse is greater than either of the two sides about the right angle.*

8. Theorem. *If oblique lines drawn from a point in a perpendicular to a line cut off unequal distances from the foot of the perpendicular, they are unequal, and the more remote is the greater; and conversely, if two oblique lines drawn from a point in a perpendicular are unequal, the greater cuts off a greater distance from the foot of the perpendicular.*

9. Theorem. *If a perpendicular is erected at the middle point of a straight line, any point not in the perpendicular is nearer that extremity of the line which is on the same side of the perpendicular.*

Corollary. *Two points equidistant from the extremities of a straight line determine a perpendicular to the line at its middle point.*

10. Theorem. *Two triangles are equal when they have three sides of one equal, respectively, to three sides of the other.*

11. Theorem. *If two lines in a plane erected perpendicular to a third are unequal, the line joining their extremities makes unequal angles with them, the greater angle with the shorter perpendicular.*

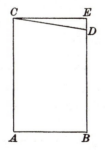

Proof. Suppose $AC > BD$. Produce BD, making $BE = AC$. Then $BEC = ACE$. But $BDC > BEC$, by (1), and ACD is a part of ACE. Therefore, all the more $BDC > ACD$.

12. Theorem. *If the two angles at C and D are equal, the perpendiculars are equal, and if the angles are unequal, the perpendiculars are unequal, and the longer perpendicular makes the smaller angle.*

13. Theorem. *If two lines are perpendicular to a third, points on either equidistant from the third are equidistant from the other.*

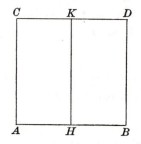

Proof. Let AB and CD be perpendicular to HK, and on CD take any two points, C and D, equidistant from K; then C and D will be equidistant from AB. For by superposition we can make D fall on C, and then DB will coincide with CA by (6).

The following propositions of Solid Geometry depend directly on the preceding and hold true at least for any restricted portion of space.

14. Theorem. *If a line is perpendicular to two intersecting lines at their intersection, it is perpendicular to all lines of their plane passing through this point.*

15. Theorem. *If two planes are perpendicular, a line drawn in one perpendicular to their intersection is perpendicular to the other, and a line drawn through any point of one perpendicular to the other lies entirely in the first.*

16. Theorem. *If a line is perpendicular to a plane, any plane through that line is perpendicular to the plane.*

17. Theorem. *If a plane is perpendicular to each of two intersecting planes, it is perpendicular to their intersection.*

III. THE THREE HYPOTHESES

The angles at the extremities of two equal perpendiculars are either right angles, acute angles, or obtuse angles, at least for restricted figures. We shall distinguish the three cases by speaking of them as the hypothesis of the right angle, the hypothesis of the acute angle, and the hypothesis of the obtuse angle, respectively.

1. Theorem. *The line joining the extremities of two equal perpendiculars is, at least for any restricted portion of the plane, equal to, greater than, or less than the line joining their feet in the three hypotheses, respectively.*

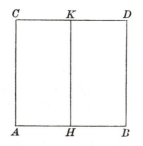

Proof. Let AC and BD be the two equal perpendiculars and HK a third perpendicular erected at the middle point of AB. Then HA and KC are perpendicular to HK, and KC is equal to, greater than, or less than HA, according as the angle at C is equal to, less than, or greater than the angle at A (II, 12). Hence, CD, the double of KC, is equal to, greater than, or less than AB in the three hypotheses, respectively.

Conversely, if CD is given equal to, greater than, or less than AB, there is established for this figure the first, second, or third hypothesis, respectively.

Corollary. *If a quadrilateral has three right angles, the sides adjacent to the fourth angle are equal to, greater than, or less than the sides opposite them, according as the fourth angle is right, acute, or obtuse.*

2. Theorem. *If the hypothesis of a right angle is true in a single case in any restricted portion of the plane, it holds true in every case and throughout the entire plane.*

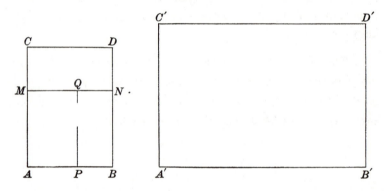

Proof. We have now a rectangle; that is, a quadrilateral with four right angles. By the corollary to the last proposition, its opposite sides are equal. Equal rectangles can be placed together so as to form a rectangle whose sides shall be any given multiples of the corresponding sides of the given rectangle.

Now let $A'B'$ be any given line and $A'C'$ and $B'D'$ two equal lines perpendicular to $A'B'$ at its extremities. Divide $A'C'$, if necessary, into a number of equal parts so that one of these parts shall be less than AC, and on AC and BD lay off AM and BN equal to one of these parts, and draw MN. $ABNM$ is a rectangle; for otherwise MN would be greater than or less than AB and CD, and the angles at M and N would all be acute angles or all obtuse angles, which is impossible, since their sum is exactly four right angles. Again, divide $A'B'$ into a sufficient number of equal parts, lay off one of these parts on AB and on MN, and form the rectangle $APQM$. Rectangles equal to this can be placed together so as exactly to cover the figure $A'B'D'C'$, which must therefore itself be a rectangle.

3. Theorem. *If the hypothesis of the acute angle or the hypothesis of the obtuse angle holds true in a single case within a restricted portion of the plane, the same hypothesis holds true for every case within any such portion of the plane.*

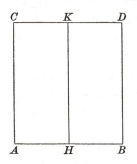

Proof. Let CD move along AC and BD, always cutting off equal distances on these two lines; or, again, let AC and BD move along on the line AB towards HK or away from HK, always remaining perpendicular to AB and their feet always at equal distances from H. The angles at C and D vary continuously and must therefore remain acute or obtuse, as the case may be, or at some point become right angles. There would then be established the hypothesis of the right angle, and the hypothesis of the acute angle or of the obtuse angle could not exist even in the single case supposed.

The angles at C and D could not become zero nor 180° in a restricted portion of the plane; for then the three lines AC, CD, and BD would be one and the same straight line.

4. Theorem. *The sum of the angles of a triangle, at least in any restricted portion of the plane, is equal to, less than, or greater than two right angles, in the three hypotheses, respectively.*

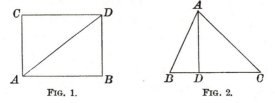

FIG. 1.   FIG. 2.

Proof. Given any right triangle, ABD (Fig. 1), with right angle at B, draw AC perpendicular to AB and equal to BD. In the triangles ADC and DAB, $AC = BD$ and AD is common, but DC is equal to, greater than, or less than AB in the three hypotheses, respectively. Therefore, DAC is equal to, greater than, or less than ADB in the three hypotheses, respectively (II, 4). Adding BAD to both of these angles, we have $ADB + BAD$ equal to, less than, or greater than the right angle BAC.

Now at least two angles of any restricted triangle are acute. The perpendicular, therefore, from the vertex of the third angle upon its opposite side will meet this side within the triangle and divide the triangle into two right triangles. Therefore, in any restricted triangle the sum of the angles is equal to, less than, or greater than two right angles in the three hypotheses, respectively.

We will call the amount by which the angle-sum of a triangle exceeds two right angles its excess. The excess of a polygon of n sides is the amount by which the sum of its angles exceeds $n - 2$ times two right angles.

It will not change the excess if we count as additional vertices any number of points on the sides, adding to the sum of the angles two right angles for each of these points.

5. Theorem. *The excess of a polygon is equal to the sum of the excesses of any system of triangles into which it may be divided.*

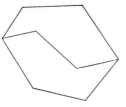

Proof. If we divide a polygon into two polygons by a straight or broken line, we may assume that the two points where it meets the boundary are vertices. If the dividing line is a broken line, broken at p points, the total sum of the angles of the two polygons so formed will be equal to the sum of the angles of the original polygon plus four right angles for each of these p points, and the sides of the two polygons will be the sides of the original polygon, together with the $p + 1$ parts into which the dividing line is separated by the p points, each part counted twice.

Let S be the sum of the angles of the original polygon, and n the number of its sides. Let S' and n', S'' and n'' have the same meanings for the two polygons into which it is divided. Then we have, writing R for right angle,

$$S' + S'' = S + 4pR,$$

and
$$n' + n'' = n + 2\,(p + 1).$$

Therefore, $\quad S' - 2\,(n' - 2)\,R + S'' - 2\,(n'' - 2)\,R$

$$= S + 4pR - 2\,(n + 2p - 2)\,R$$

$$= S - 2\,(n - 2)\,R.$$

Any system of triangles into which a polygon may be divided is produced by a sufficient number of repetitions of the above process. Always the excess of the polygon is equal to the sum of the excesses of the parts into which it is divided.

We may extend the notion of excess and apply it to any combination of different portions of the plane bounded completely by straight lines.

Instead of considering the sum of the angles of a polygon, we may take the sum of the exterior angles. The amount by which this sum falls short of four right angles equals the excess of the polygon. We may speak of it as the deficiency of the exterior angles.

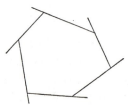

The sum of the exterior angles is the amount by which we turn in going completely around the figure, turning at each vertex from one side to the next. If we are considering a combination of two or more polygons, we must traverse the entire boundary and so as always to have the area considered on one side, say on the left.

6. Theorem. *The excess of polygons is always zero, always negative, or always positive.*

Proof. We know that this theorem is true of restricted triangles, but any finite polygon may be divided into a finite number of such triangles, and by the last theorem the excess of the polygon is equal to the sum of the excesses of the triangles.

When the excess is negative, we may call it deficiency, or speak of the excess of the exterior angles.

Corollary. *The excess of a polygon is numerically greater than the excess of any part which may be cut off from it by straight lines, except in the first hypothesis, when it is zero.*

The following theorems apply to the second and third hypotheses.

7. Theorem. *By diminishing the sides of a triangle, or even one side while the other two remain less than some fixed length, we can diminish its area indefinitely, and the sum of its angles will approach two right angles as limit.*

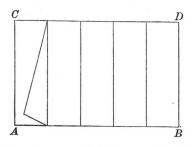

Proof. Let $ABDC$ be a quadrilateral with three right angles, A, B, and C. A perpendicular moving along AB will constantly increase or decrease; for if it could increase a part of the way and decrease a part of the way there would be different positions where the perpendiculars have the same length; a perpendicular midway between them would be perpendicular to CD also, and we should have a rectangle.

Divide AB into n equal parts, and draw perpendiculars through the points of division. The quadrilateral is divided into n smaller quadrilaterals, which can be applied one to another, having a side and two adjacent right angles the same in all. Beginning at the end where the perpendicular is the shortest, each quadrilateral can be placed entirely within the next. Therefore, the first has its area less than $\frac{1}{n}$th of the area of the original quadrilateral, and its deficiency or excess less than $\frac{1}{n}$th of the deficiency or excess of the whole. Now any triangle whose sides are all less than AC or BD, and one of whose sides is less than one of the subdivisions of AB, can be placed entirely within this smallest quadrilateral. Such a triangle has its area and its deficiency or excess less than $\frac{1}{n}$ th of the area and of the deficiency or excess of the original quadrilateral.

Thus, a triangle has its area and deficiency or excess less than any assigned area and deficiency or excess, however small, if at least one side is taken sufficiently small, the other two sides not being indefinitely large.

8. Theorem. *Two triangles having the same deficiency or excess have the same area.*

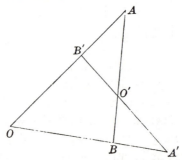

Proof. Let AOB and $A'OB'$ have the same deficiency or excess and an angle of one equal to an angle of the other. If we place them together so that the equal angles coincide, the triangles will coincide and be entirely equal, or there will be a quadrilateral common to the two, and, besides this, two smaller triangles having an angle the same in both and the same deficiency or excess. Putting these together, we find again a quadrilateral common to both and a third pair of triangles having an angle the same in both and the same deficiency or excess. We may continue this process indefinitely, unless we come to a pair of triangles which coincide; for at no time can one triangle of a pair be contained entirely within the other, since they have the same deficiency or excess.

Let *so* denote the sum of the sides opposite the equal angles of the first two triangles, *sa* the sum of the adjacent sides, and *s'a* that portion of the adjacent sides counted twice, which is common to the two triangles when they are placed together. Writing o' and a' for the second pair of triangles, o'' and a'' for the third pair, etc., we have

$$sa \ = s'a \ + so', \qquad\qquad so \ = sa',$$
$$sa' = s'a' + so'', \qquad\qquad so' = sa'',$$
$$sa'' = s'a'' + so''', \text{ etc.} \qquad so'' = sa''', \text{ etc.}$$
$$\therefore sa \ = s'a \ + s'a'' + s'a^{\text{iv}} + \cdots,$$
$$sa' = s'a' + s'a''' + s'a^{\text{v}} + \cdots.$$

Therefore, the expressions $s'a, s'a', s'a'', \cdots$ diminish indefinitely. Each of these is made up of a side counted twice from one and a side counted twice from the other of a pair of triangles. Thus, if we carry the process sufficiently far, the remaining triangles can be made to have at least one side as small as we please, while all the sides diminish and are less, for example, than the longest of the sides of the original triangles. Therefore, the areas of the remaining triangles diminish indefinitely, and as the difference of the areas remains the same for each pair of triangles, this difference must be zero. The triangles of each pair and, in particular, the first two triangles have the same area.

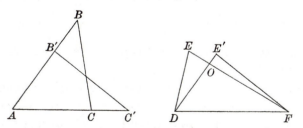

Let ABC and DEF have the same deficiency or excess, and suppose $AC < DF$. Produce AC to C', making $AC' = DF$.

Then there is some point, B', on AB between A and B such that $AB'C'$ has the same deficiency or excess and the same area as ABC. Place $AB'C'$ upon DEF so that AC' will coincide with DF, and let $DE'F$ be the position which it takes. If the triangles do not coincide, the vertex of each opposite the common side DF lies outside of the other. The two triangles have in common a triangle, say DOF, and besides this there remain of the two triangles two smaller triangles which have one angle the same in both and the same deficiency or excess. These two triangles, and therefore the original triangles, have the same area.

9. Theorem. *The areas of any two triangles are proportional to their deficiencies or excesses.*

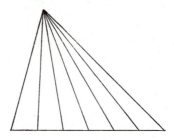

Proof. A triangle may be divided into n smaller triangles having equal deficiencies or excesses and equal areas by lines drawn from one vertex to points of the opposite side. Each of these triangles has for its deficiency or excess $\frac{1}{n}$ th of the deficiency or excess of the original triangle, and for its area $\frac{1}{n}$ th of the area of the original triangle.

When the deficiencies or excesses of two triangles are commensurable, say in the ratio $m:n$, we can divide them into m and n smaller triangles, respectively, all having the same deficiency or excess and the same area. The areas of the given triangles will therefore be in the same ratio, $m:n$.

When the deficiencies or excesses of two triangles, A and B, are not commensurable, we may divide one triangle, A, as above, into any number of equivalent parts, and take parts equivalent to one of these as many times as possible from the

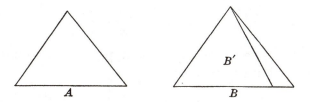

other, leaving a remainder which has a deficiency or excess less than the deficiency or excess of one of these parts. The portion taken from the second triangle forms a triangle, B'. A and B' have their areas proportional to their deficiencies or excesses, these being commensurable. Now increase indefinitely the number of parts into which A is divided. These parts will diminish indefinitely, and the remainder when we take B' from B will diminish indefinitely. The deficiency or excess and the area of B' will approach those of B, and the triangles A and B have their areas and their deficiencies or excesses proportional.

Corollary. *The areas of two polygons are to each other as their deficiencies or excesses.*

10. Theorem. *Given a right triangle with a fixed angle; if the sides of the triangle diminish indefinitely, the ratio of the opposite side to the hypothenuse and the ratio of the adjacent side to the hypothenuse approach as limits the sine and cosine of this angle.*

Proof. Lay off on the hypothenuse any number of equal lengths. Through the points of division A_1, A_2, $\cdots$ draw perpendiculars A_1C_1, A_2C_2, $\cdots$ to the base, and to these lines produced draw perpendiculars A_2D_1, A_3D_2, $\cdots$ each from the next point of division of the hypothenuse.

The triangles OA_1C_1 and $A_2A_1D_1$ are equal (II, 6, Cor.).

$$C_2A_2 \gtrless C_1D_1 \quad \text{and} \quad C_1C_2 \lessgtr D_1A_2;$$

therefore, $$\frac{C_2A_2}{OA_2} \gtrless \frac{C_1A_1}{OA_1} \quad \text{and} \quad \frac{OC_2}{OA_2} \lessgtr \frac{OC_1}{OA_1},$$

the upper sign being for the second hypothesis and the lower sign for the third hypothesis.

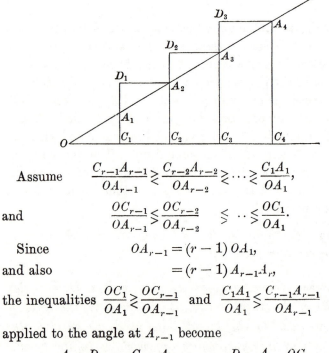

Assume $$\frac{C_{r-1}A_{r-1}}{OA_{r-1}} \gtrless \frac{C_{r-2}A_{r-2}}{OA_{r-2}} \gtrless \cdots \gtrless \frac{C_1A_1}{OA_1},$$

and $$\frac{OC_{r-1}}{OA_{r-1}} \lessgtr \frac{OC_{r-2}}{OA_{r-2}} \lessgtr \cdots \lessgtr \frac{OC_1}{OA_1}.$$

Since $$OA_{r-1} = (r-1)\,OA_1,$$

and also $$= (r-1)\,A_{r-1}A_r,$$

the inequalities $\dfrac{OC_1}{OA_1} \gtrless \dfrac{OC_{r-1}}{OA_{r-1}}$ and $\dfrac{C_1A_1}{OA_1} \lessgtr \dfrac{C_{r-1}A_{r-1}}{OA_{r-1}}$

applied to the angle at A_{r-1} become

$$\frac{A_{r-1}D_{r-1}}{A_{r-1}A_r} \gtrless \frac{C_{r-1}A_{r-1}}{OA_{r-1}} \quad \text{and} \quad \frac{D_{r-1}A_r}{A_{r-1}A_r} \lessgtr \frac{OC_{r-1}}{OA_{r-1}}.$$

The first of these two inequalities may be written

$$\frac{A_{r-1}D_{r-1}}{C_{r-1}A_{r-1}} \gtrless \frac{A_{r-1}A_r}{OA_{r-1}}.$$

Add 1 to both members,

$$\frac{C_{r-1}D_{r-1}}{C_{r-1}A_{r-1}} \gtrless \frac{OA_r}{OA_{r-1}},$$

or

$$\frac{C_{r-1}D_{r-1}}{OA_r} \gtrless \frac{C_{r-1}A_{r-1}}{OA_{r-1}}.$$

But

$$C_rA_r \gtrless C_{r-1}D_{r-1}.$$

∴

$$\frac{C_rA_r}{OA_r} \gtrless \frac{C_{r-1}A_{r-1}}{OA_{r-1}}.$$

Again,

$$C_{r-1}C_r \lessgtr D_{r-1}A_r.$$

Hence, from the second inequality above, we have

$$\frac{C_{r-1}C_r}{A_{r-1}A_r} \lessgtr \frac{OC_{r-1}}{OA_{r-1}},$$

or

$$\frac{C_{r-1}C_r}{OC_{r-1}} \lessgtr \frac{A_{r-1}A_r}{OA_{r-1}}.$$

Add 1 to both members,

$$\frac{OC_r}{OC_{r-1}} \lessgtr \frac{OA_r}{OA_{r-1}},$$

or

$$\frac{OC_r}{OA_r} \lessgtr \frac{OC_{r-1}}{OA_{r-1}}.$$

The ratios $\frac{CA}{OA}$ and $\frac{OC}{OA}$ being less than 1, and always increasing or always decreasing when the hypothenuse decreases, approach definite limits. These limits are continuous functions of A; if we vary the angle of any right triangle continuously, keeping the hypothenuse some fixed length, the other two sides will vary continuously, and the limits of their ratios to the hypothenuse must, therefore, vary continuously.

Calling the limits for the moment sA and cA, we may extend their definition, as in Trigonometry, to any angles, and prove that all the formulæ of the sine and cosine hold for these functions. Then for certain angles, 30°, 45°, 60°, we can prove that they have the same values as the sine and cosine, and their values for all other angles as determined from their values for these angles will be the same as the corresponding values of the sine and cosine.

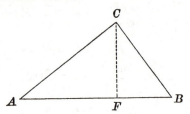

Draw a perpendicular, CF, from the right angle C to the hypothenuse AB. The angle FCB is not equal to A, but the difference, being proportional to the difference of areas of the two triangles ABC and FBC, diminishes indefinitely when the sides of the triangles diminish. From the relation

$$\frac{AF}{AC}\frac{AC}{AB} + \frac{FB}{BC}\frac{BC}{AB} = 1,$$

we have, by passing to the limit,

$$(cA)^2 + (sA)^2 = 1.$$

Let x and y be any two acute angles, and draw the figures used to prove the formulæ for the sine and cosine of the sum of two angles.

The angles x and y remaining fixed, we can imagine all of the lines to decrease indefinitely, and the functions sx, cx, sy, etc., are the limits of certain ratios of these lines.

$$\frac{CA}{OA} = \frac{CE}{OB}\frac{OB}{OA} + \frac{EA}{BA}\frac{BA}{OA},$$

$$\pm \frac{OC}{O.1} = \frac{OD}{OB}\frac{OB}{OA} - \frac{CD}{BA}\frac{BA}{OA}$$

$$\left(-\frac{OC}{OA} \text{ in the second figure}\right).$$

The angles at M are equal in the two triangles EMB and CMO, and we may write

$$\frac{CM}{OM} = \frac{ME + \delta}{MB} = \frac{ME \pm CM + \delta}{MB \pm OM},$$

where δ has the limit zero.

$$\therefore \lim \frac{CE}{OB} = \lim \frac{CM}{OM} = sx.$$

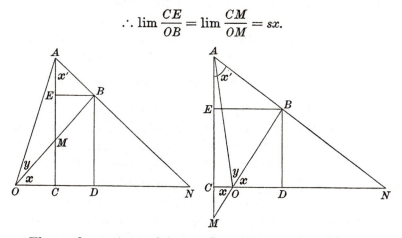

The angle EAB, or x', is not the same as x, but differs from x only by an amount which is proportional to the difference of the areas of the triangles OMC and MAB, and which, therefore, diminishes indefinitely. Thus, the limits of sx' and cx' are sx and cx.

Finally, as the two triangles ACN and BDN have the angle N in common, we may write

$$\frac{DN}{BN} = \frac{CN + \delta'}{AN} = \frac{CN - DN + \delta'}{AN - BN},$$

where the limit of δ' is zero.

$$\therefore \lim \frac{CD}{AB} = \lim \frac{CN}{AN} = sx.$$

Now at the limits our identities become

$$s(x + y) = sx \cdot cy + cx \cdot sy,$$
$$c(x + y) = cx \cdot cy - sx \cdot sy.$$

By induction, these formulæ are proved true for any angles. Other formulæ sufficient for calculating the values of these functions from their values for 30°, 45°, and 60° are obtained from these two by algebraic processes.

If the sides of an isosceles right triangle diminish indefinitely, the angle does not remain fixed but approaches 45°, and the ratios of the two sides to the hypothenuse approach as limits $s\,45°$ and $c\,45°$. Therefore, these latter are equal, and since the sum of their squares is 1, the value of each is $\dfrac{1}{\sqrt{2}}$, the same as the value of the sine and cosine of 45°.

Again, bisect an equilateral triangle and form a triangle in which the hypothenuse is twice one of the sides. When the sides diminish, preserving this relation, the angles approach 30° and 60°. Therefore, the functions, s and c, of these angles have values which are the same as the corresponding values of the sine and cosine of the same angles.

Corollary. *When any plane triangle diminishes indefinitely, the relations of the sides and angles approach those of the sides and angles of plane triangles in the ordinary geometry and trigonometry with which we are familiar.*

11. Theorem. *Spherical geometry is the same in the three hypotheses, and the formulæ of spherical trigonometry are exactly those of the ordinary spherical trigonometry.*

Proof. On a sphere, arcs of great circles are proportional to the angles which they subtend at the centre, and angles on a sphere are the same as the diedral angles formed by the planes of the great circles which are the sides of the angles. Their relations are established by drawing certain plane triangles which may be made as small as we please, and therefore may be assumed to be like the plane triangles in the hypothesis of a right angle. These relations are, therefore, those of the ordinary Spherical Trigonometry.

The three hypotheses give rise to three systems of Geometry, which are called the Parabolic, the Hyperbolic, and the Elliptic Geometries. They are also called the Geometries of Euclid, of Lobachevsky, and of Riemann. The following considerations exhibit some of their chief characteristics.

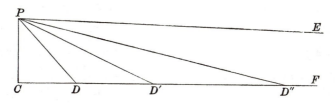

Given PC perpendicular to a line, CF; on the latter we take

$$CD = PC,$$
$$DD' = PD,$$
$$D'D'' = PD', \text{ etc.}$$

Now if PC is sufficiently short (restricted), it is shorter than *any* other line from P to the line CF; for any line as short as PC or shorter would be included in a restricted portion of the plane about the point P, for which the perpendicular is the shortest distance from the point to the line.

Therefore, $PD > PC$, $\therefore CD' > 2\,CD$,
$PD' > PC$, etc.; $CD'' > 3\,CD$, etc.

Again, in the three hypotheses, respectively,

$$CPD \lesseqqgtr \frac{\pi}{4}, \qquad \text{and} \qquad CDP \lesseqqgtr \frac{\pi}{4},$$
$$DPD' \lesseqqgtr \tfrac{1}{2}CPD, \qquad\qquad CD'P \lesseqqgtr \tfrac{1}{2}CDP,$$
$$D'PD'' \lesseqqgtr \tfrac{1}{2}DPD', \text{ etc.}, \qquad CD''P \lesseqqgtr \tfrac{1}{2}CD'P, \text{ etc.}$$

At P we have a series of angles. In the first hypothesis there is an infinite number of these angles, and the series forms a geometrical progression of ratio $\frac{1}{2}$, whose value is exactly $\frac{\pi}{2}$. In the second hypothesis there is also an infinite number of these angles, and the terms of the series are less than the terms of the geometrical progression. The value of the series is, therefore, less than $\frac{\pi}{2}$. In the third hypothesis we have a series whose terms are greater than those of the geometrical progression, and, therefore, whether the series is convergent or divergent, we can get more than $\frac{\pi}{2}$ by taking a sufficient number of terms. In other words, we can get a right angle or more than a right angle at P by repeating this process a certain finite number of times.

The angles at D, D', D'', $\cdots$ are exactly equal to the terms of the series of angles at P. In the first two hypotheses they approach zero as a limit.

The distances CD, CD', CD'', $\cdots$ increase each time by more than a definite quantity, CD; therefore, if we repeat the process an unlimited number of times, these distances will increase beyond all limit. Thus, in the first and second hypotheses we prove that a straight line must be of infinite length.

In the hypothesis of the obtuse angle the line perpendicular to PC at the point P will intersect CF in a point at a certain finite distance from C, one of the D's, or some point between. On the other side of PC this same perpendicular will intersect FC produced at the same distance. But we have assumed that two different straight lines cannot intersect in two points; therefore, for us the third hypothesis cannot be true unless the straight line is of finite length returning into itself, and these two points are one and the same point, its distance from C in either direction being one-half the entire length of the line. In this way, however, we can build up a consistent Geometry on the third hypothesis, and this Geometry it is which is called the Elliptic Geometry.

The constructions would have been the same, and very nearly all the statements would have been the same, if we had taken CD any arbitrary length on CF.

The restriction which we have placed upon some of the propositions of this chapter is necessary in the third hypothesis.

Thus, in the proof that the exterior angle of a triangle is greater than the opposite interior angle, the line AD drawn through the vertex A to the middle point D of the opposite

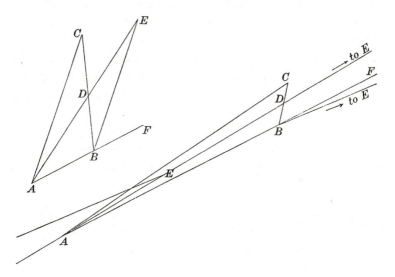

side was produced so as to make $AE = 2\,AD$. If AD were greater than half the entire length of the straight line determined by A and D, this would bring the point E past the point A, and the angle CBE, which is equal to the angle C, instead of being a part of the exterior angle CBF, becomes greater than this exterior angle.

Again, if two angles of a triangle are equal and the side between them is just an entire straight line, it does not follow necessarily that the opposite sides are equal. It may be said, however, that the opposite sides form one continuous line, and, therefore, this figure is not strictly a triangle, but a figure

somewhat like a lune. The points A and B are the same point, and the angles A and B are vertical angles.

Finally, though we assume that the shortest path between two points is a straight line, it is not always true that a straight line drawn between two points is the shortest path between them. We can pass from one point to another in two ways on a straight line; namely, over each of the two parts into which the two points divide the line determined by them. One of these parts will usually be shorter than the other, and the longer part will be longer than some paths along broken lines or curved lines.

When, however, the straight line is of infinite length, that is, in the hypothesis of the right angle and in the hypothesis of the acute angle, all the propositions of this chapter hold without restriction.

The Euclidean Geometry is familiar to all. We will now make a detailed study of the Geometry of Lobachevsky, and then take up in the same way the Elliptic Geometry.

CHAPTER II

THE HYPERBOLIC GEOMETRY

We have now the hypothesis of the acute angle. Two lines in a plane perpendicular to a third diverge on either side of their common perpendicular. The sum of the angles of a triangle is less than two right angles, and the propositions of the last chapter hold without restriction.

I. PARALLEL LINES

From any point, P, draw a perpendicular, PC, to a given line, AB, and let PD be any other line from P meeting CB in D. If D move off indefinitely on CB, the line PD will approach a limiting position PE.

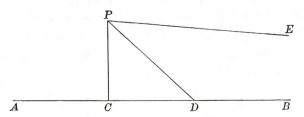

PE is said to be parallel to CB at P. PE makes with PC an angle, CPE, which is called the angle of parallelism for the perpendicular distance PC. It is less than a right angle by an amount which is the limit of the deficiency of the triangle PCD. On the other side of PC we can find another line parallel to CA and making with PC the same angle of parallelism. We say that PE is parallel to AB towards that part which is on the same side of PC with PE. Thus, at any point there are two parallels to a line, but only one towards one part of the line. Lines through P which make with PC an angle greater than the angle of parallelism and less than its supplement do not meet AB at all. We write $\Pi(p)$ to denote the angle of parallelism for a perpendicular distance, p.

1. Theorem. *A straight line maintains its parallelism at all points.*

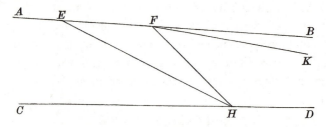

Let AB be parallel to CD at E and let F be any other point of AB on either side of E, to prove that AB is parallel to CD at F.

Proof. To H, on CD, draw EH and FH. If H move off indefinitely on CD, these two lines will approach positions of parallelism with CD. But the limiting position of EH is the line AB passing through F, and if the limiting position of FH were some other line, FK, F would be the limiting position of H, the intersection of EH and FH.

2. Theorem. *If one line is parallel to another, the second is parallel to the first.*

Given AB parallel to CD, to prove that CD is parallel to AB.

Proof. Draw AC perpendicular to CD. The angle CAB will be acute; therefore, the perpendicular CE from C to AB must fall on that side of A towards which the line AB is parallel to CD (Chap. I, II, 1). The angle ECD is then acute and less than CEB, which is a right angle. That is, we have

$$CAB < ACD, \quad \text{and} \quad CEB > ECD.$$

If the line CE revolve about the point C to the position of CA, the angle at E will decrease to the angle A, and the angle at C will increase to a right angle. There will be some position, say CF, where these two angles become equal; that is,

$$CFB = FCD.$$

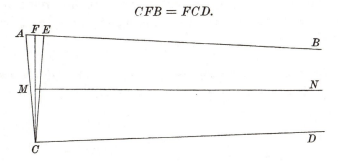

Draw MN perpendicular to CF at its middle point and revolve the figure about MN as an axis. CD will fall upon the original position of AB, and AB will fall upon the original position of CD. Therefore, CD is parallel to AB.

Corollary. *FB and CD are both parallel to MN.*

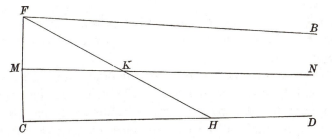

Proof. FB and CD are symmetrically situated with respect to MN, and cannot intersect MN since they do not intersect each other. Draw FH to H, on CD, intersecting MN in K. If H move off indefinitely on CD, FH will approach the position of FB as a limit. Now K cannot move off indefinitely before H does, for $FK < FH$. But again, when H moves off indefinitely, K cannot approach some limiting position at a finite distance on MN; for FB, and therefore CD, would then intersect MN and each other at this point. Therefore, H and K must move off together, and the limiting position of FH must be at the same time parallel to CD and MN.

In the same way we can prove that any line lying in a plane between two parallels must intersect one of them or be parallel to both.

3. Theorem. *Two lines parallel to a third towards the same part of the third are parallel to each other.*

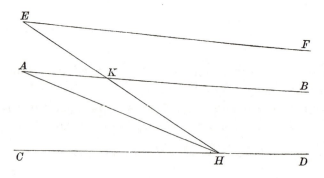

First, when they are all in the same plane.

Let AB and EF be parallel to CD, to prove that they are parallel to each other.

Proof. Suppose AB lies between the other two. To H, any point on CD, draw AH and EH, and let K be the point where EH intersects AB. As H moves off indefinitely on CD, AH and EH approach as limiting positions AB and EF. Now K cannot move off indefinitely before H does, for $EK < EH$. But again, when H moves off indefinitely, K cannot approach some limiting position at a finite distance on AB; for this point would be the intersection of AB and EF, and the limiting position of H, whereas H moves off indefinitely on CD. Therefore, H and K must move off together, and the limiting position of EH must be at the same time parallel to CD and AB.

If *AB*, lying between the other two, is given parallel to *CD* and *EF*, *EF* must be parallel to *CD*; for a line through *E* parallel to *CD* would be parallel to *AB*, and only one line can be drawn through *E* parallel to *AB* towards the same part.

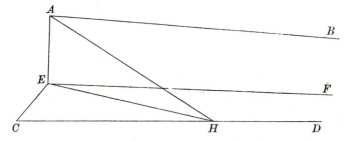

Second, when the lines are not all in the same plane.

Let *AB* and *CD* be two parallel lines and let *E* be any point not in their plane.

Proof. To *H* on *CD* draw *AH* and *EH*. As *H* moves off indefinitely, *AH* approaches the position of *AB*, and the plane *EAH* the position of the plane *EAB*. Therefore, the limiting position of *EH* is the intersection of the planes *ECD* and *EAB*. The intersection of these planes is, then, parallel to *CD*, and in the same way we prove that it is parallel to *AB*.

Now, if *EF* is given as parallel to one of these two lines towards the part towards which they are parallel, it must be the intersection of the two planes determined by them and the point *E*, and therefore parallel to the other line also.

4. Theorem. *Parallel lines continually approach each other.*

Let *AB* and *CD* be parallel, and from *A* and *B*, any points on *AB*, drop perpendiculars *AC* and *BD* to *CD*. Supposing that *B* lies beyond *A* in the direction of parallelism, we are to prove that *BD* < *AC*.

Proof. At *H*, the middle point of *CD*, erect a perpendicular meeting *AB* in *K*. The angle *BKH* is an acute angle, and the angle *AKH* is an obtuse angle.   Therefore, a perpendicular to *HK* at *K* must meet *CA* in some point, *E*, between *C* and *A*

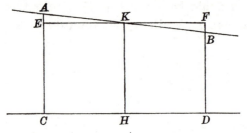

and *DB* produced in some point, *F*, beyond *B*.   But *DF = CE* (Chap. I, I, 12); therefore, *DB < CA*.

Corollary. *If AB and CD are parallel and AC makes equal angles with them (like FC in 2 above), then EF, cutting off equal distances on these two lines, AE = CF, on the side towards which they are parallel, will be shorter than AC,*

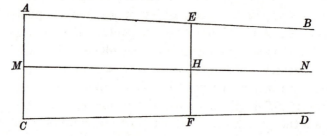

Proof. *MN*, perpendicular to *AC* at its middle point, is parallel to *AB* and bisects *EF*, the figure being symmetrical with respect to *MN*.   *EH*, the half of *EF*, is less than *AM*, and therefore *EF* is less than *AC*.

5. Theorem. *As the perpendicular distance varies, starting from zero and increasing indefinitely, the angle of parallelism decreases from a right angle to zero.*

Proof. In the first place the angle of parallelism, which is acute as long as the perpendicular distance is positive, will be made to differ from a right angle by less than any assigned value if we take a perpendicular distance sufficiently small.

For, ADE being any given angle as near a right angle as we please, we can take a point, L, on DE and draw LR perpendicular to DA at R. The angle RDL must increase to become the angle of parallelism for the perpendicular distance RD.

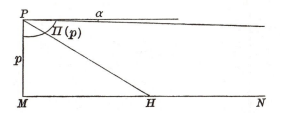

Now let p be the length of a given perpendicular PM, and let α be the amount by which its angle of parallelism differs from $\frac{\pi}{2}$; that is, say

$$\Pi(p) = \frac{\pi}{2} - \alpha.$$

PM, being perpendicular to MN, and H any point on MN, the angle MPH approaches as a limit the angle of parallelism, $\Pi(p)$, when H moves off indefinitely on MN. The line PH meets the line MN as long as $MPH < \Pi(p)$, and by taking MPH sufficiently near $\Pi(p)$, but less, we can make the angle MHP as small as we please (see p. 27).

In figure on page 38, let AC be perpendicular to AB, D being any point on AC and DE parallel to AB. Draw DK beyond DE, making with DE an angle, $EDK = \Pi(p)$, and make $DK = p$. TF, perpendicular to DK at K, will be parallel to DE and AB.

By placing *PMN* of the last figure upon *DKT*, we see that *DC* will meet *KT* in a point, *G* if

$$KDC < \Pi(p),$$

that is, if
$$ADE > 2\,\alpha.$$

Then in the right triangle *DKG*,

$$DGK + KDG < \frac{\pi}{2}.$$

But
$$ADE + KDG = \frac{\pi}{2} + \alpha;$$

therefore,
$$DGK < ADE - \alpha.$$

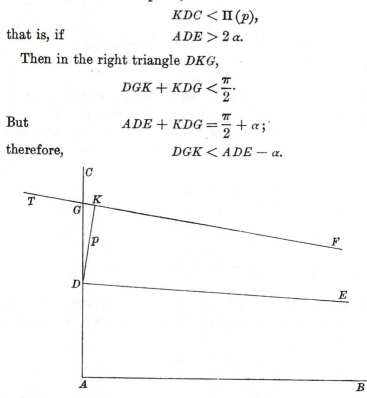

Starting from the point *G*, we can repeat this construction, and each time we subtract from the angle of parallelism an amount greater than *α*. We can continue this process until the angle of parallelism becomes equal to or less than 2 *α*.

If the point *D* move along *AC*, *DE* remaining constantly parallel to *AB*, the angle at *D* will constantly diminish, and by letting *D* move sufficiently far on *AC* we can reach a point where this angle becomes equal to or less than 2 *α*.

Suppose *D* is at the point where the angle of parallelism is just 2 *α*. Then, if we draw *DK* and *TF* as before, *KT* will be parallel to DC.   All the parallels to AB lying between AB and this position of TF meet AC, and as the parallel moves towards this position of TF, the angle of parallelism at D approaches zero, and the point D moves off indefinitely.

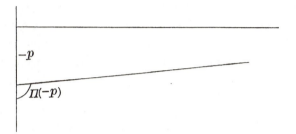

For an obtuse angle we may take p negative, and we have

$$\Pi\,(-p) = \pi - \Pi\,(p).$$

6. Theorem. *The perpendiculars erected at the middle points of the sides of a triangle are all parallel if two of them are parallel.*

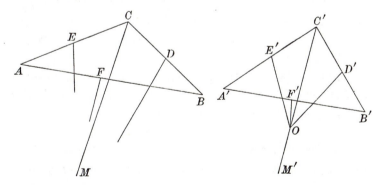

Let A, B, and C be the vertices of the triangle, and D, E, and F, respectively, the middle points of the opposite sides. Suppose the perpendiculars at D and E are given parallel, to prove that the perpendicular at F is parallel to them.

Proof. Draw CM through C parallel to the two given parallel perpendiculars. CM forms with the two sides at C angles of parallelism $\Pi\left(\dfrac{a}{2}\right)$ and $\Pi\left(\dfrac{b}{2}\right)$, of which the angle at C is the sum or difference according as C lies between the given perpendiculars or on the same side of both. By properly diminishing these angles at C, keeping the lengths of CA and CB unchanged, we can make the perpendiculars at their middle points D and E intersect CM, and therefore each other, at any distance from C greater than $\dfrac{a}{2}$ and greater than $\dfrac{b}{2}$.

Let $A'B'C'$ be the triangle so formed, O the point where the two given perpendiculars meet, and $C'M'$ the line through O. In the triangle $A'B'C'$, the three perpendiculars meet at the point O (Chap. I, I, 5). Now we can let O move off on $C'M'$, the construction remaining the same. That is, we let the lines $C'A'$ and $C'B'$ rotate about C' without changing their lengths, in such a manner that the three perpendiculars $D'O$, $E'O$, and $F'O$ shall always pass through O. As O moves off indefinitely, the angles at C' approach $\Pi\left(\dfrac{a}{2}\right)$ and $\Pi\left(\dfrac{b}{2}\right)$ as limits, and the three perpendiculars approach positions of parallelism with $C'M'$ and with each other. But the triangle $A'B'C'$ approaches as a limit a triangle which is equal to ABC, having two sides and the included angle equal, respectively, to the corresponding parts of the latter. Therefore, in ABC the three perpendiculars are all parallel.

7. Theorem. *Lines which do not intersect and are not parallel have one and only one common perpendicular.*

Proof. Let AB and CD be the two lines, and from A, any point of AB, drop AC perpendicular to CD. If AC is not itself the common perpendicular, one of the angles which it makes with AB will be acute. Let this angle be on the side towards AB, so that $BAC < \dfrac{\pi}{2}$. Draw AE parallel to CD on this same side of AC. The angle EAC is less than BAC, since AB is not parallel to CD and does not intersect it. Let AH be any line drawn in the angle EAC, intersecting CD at H. If H, starting from the position of C, move off indefinitely

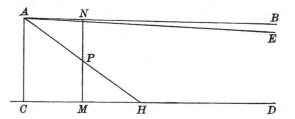

on the line CD, the angle BAH will decrease from the magnitude of the angle BAC to the angle BAE. The angle AHC will decrease *indefinitely* from the magnitude of the angle at C, which is a right angle and greater than BAC. There will be some position for which $BAH = AHC$. In this position the line NM through the middle point of AH perpendicular to one of the two given lines will be perpendicular to the other, as proved in Chap. I, I, 10.

If there were two common perpendiculars we should have a rectangle, which is impossible in the Hyperbolic Geometry.

8. Theorem. *If the perpendiculars erected at the middle*

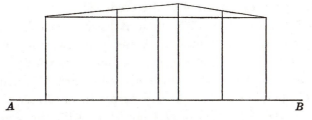

points of the sides of a triangle do not meet and are not parallel, they are all perpendicular to a certain line.

Proof. We can draw a line, AB, that will be perpendicular to two of these lines, and the perpendiculars from the three vertices of the triangle upon this line will be equal, by Chap. I, II, 13. A perpendicular to AB erected midway between any two of these three is perpendicular to the corresponding side of the triangle at its middle point (Chap. I, I, 11). Thus, all three of the perpendiculars erected at the middle points of the sides of the triangle are perpendicular to AB.

A line is parallel to a plane if it is parallel to its projection on the plane.

9. Theorem. *A line may be drawn perpendicular to a plane and parallel to any line not in the plane.*

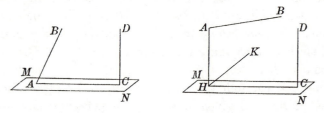

Proof. Let AB be the given line and MN the plane. If AB meets the plane MN at a point, A, we take on its projection a length, AC, such that the angle at A equals $\Pi(AC)$. Then CD, perpendicular to the plane at C, will be parallel to AB. In the same way, on the other side of the plane a perpendicular can be drawn parallel to BA produced.

If AB does not meet MN, then at least in one direction it diverges from MN. Through H, any point of the projection of AB on the plane, we can draw a line, HK, parallel to AB towards that part of AB which diverges from MN, and then draw CD parallel to this line and perpendicular to the plane.

Unless AB is parallel to MN it will meet the plane at some point, or the plane and line will have a common perpendicular, and the line will diverge from the plane in both directions.

In the latter case there are two perpendiculars that are parallel to the line, one parallel towards each part of the line.

Two perpendiculars cannot be parallel towards the same part of a line; for then they would be parallel to each other, and two lines cannot be perpendicular to a plane and parallel to each other.

II. BOUNDARY–CURVES AND SURFACES, AND EQUI-DISTANT–CURVES AND SURFACES

Having given the line AB, at its extremity, A, we take any arbitrary angle and produce the side AC so that the perpendicular erected at its middle point shall be parallel to AB. The locus of the point C is a curve which is called oricycle, or boundary-curve. AB is its axis.

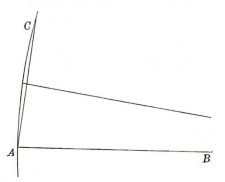

From their definition it follows that all boundary-curves are equal, and the boundary-curve is symmetrical with respect to its axis; if revolved through two right angles about its axis, it will coincide with itself.

1. Theorem. *Any line parallel to the axis of a boundary-curve may be taken for axis.*

Let AB be the axis and CD any line parallel to AB, to prove that CD may be taken as axis.

Proof. Draw AC; also to E, any other point on the curve, draw AE and CE. The perpendiculars erected at the middle points of AC and of AE are parallel to AB and CD and to each

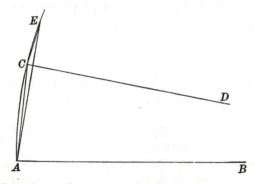

other. Therefore, the perpendicular erected at the middle point of CE, the third side of the triangle ACE, is parallel to them and to CD. CD then may be taken as axis.

Corollary. *The boundary-curve may be slid along on itself without altering its shape ; that is, it has a constant curvature.*

2. Theorem. *Two boundary-curves having a common set of axes cut off the same distance on each of the axes, and the ratio of corresponding arcs depends only on this distance.*

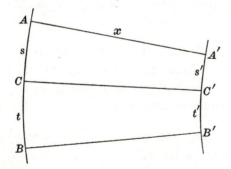

Proof. Take any two axes and a third axis bisecting the arc which the first two intercept on one of the two boundary-curves. By revolving the figure about this axis we show that the curves cut off equal distances on the two axes.

Let AA', BB', and CC' be any three axes of the two boundary-curves AB and $A'B'$; let their common length be x and let them intercept arcs s and t on AB, s' and t' on $A'B'$.

When $s = t$, $s' = t'$, and, in general,

$$\frac{s}{t} = \frac{s'}{t'},$$

as we prove, first when s and t are commensurable, and then by the method of limits when they are incommensurable. The ratio $\frac{s}{s'}$ is, therefore, a constant for the given value of x.

Write $$\frac{s}{s'} = f(x).$$

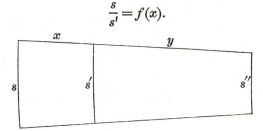

From three boundary-curves having the same set of axes, we find $$f(x + y) = f(x)f(y).$$

This property is characteristic of the exponential function whose general form is $f(x) = e^{ax}$.* Therefore, $\frac{s}{s'} = e^{ax}$, the value of a depending on the unit of measure (see below p. 76).

* Putting $y = x, 2x, \cdots (n-1)x$ in succession, we find
$$f(nx) = [f(x)]^n$$
for positive integer values of n, x being any positive quantity.

Now $$f\left(\frac{r}{s}x\right) = \left[f\left(\frac{x}{s}\right)\right]^r,$$

and this is the rth power of the sth root of the first member of the equation

$$\left[f\left(\frac{x}{s}\right)\right]^s = f(x); \qquad \therefore f\left(\frac{r}{s}x\right) = [f(x)]^{\frac{r}{s}}.$$

3. Theorem. *The area enclosed by two boundary-curves having the same axes and by two of their common axes is proportional to the difference of the intercepted arcs.*

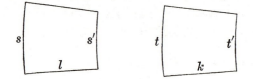

Proof. Let s and s' be the lengths of the intercepted arcs, and l the distance measured on an axis between them. Let t, t', and k be the corresponding quantities for a second figure constructed in the same way.

If the corresponding lines in the two figures are all equal, the areas are equal, for they can be made to coincide. If only $k = l$, the areas are to each other as corresponding arcs, say as $s' : t'$, proved first when the arcs are commensurable, and then by the method of limits when they are incommensurable.

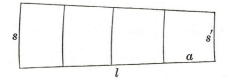

When l and k are commensurable, suppose

$$\frac{l}{m} = \frac{k}{n} = a.$$

Thus, assuming that $f(x)$ is a continuous function of x, we have proved that for all real positive values of x and n

$$f(nx) = [f(x)]^n,$$

and if we put x for n and 1 for x, we have

$$f(x) = [f(1)]^x.$$

We will write $f(1) = e^a$; then

$$f(x) = e^{ax}.$$

We can draw a series of boundary-curves at distances equal to a on the axes and divide the areas into m and n parts, respectively. If r is the ratio of arcs corresponding to the distance a, these parts will be proportional to the quantities

$$s', \ s'r, \ s'r^2, \ \cdots \ s'r^{m-1};$$

$$t', \ t'r, \ t'r^2, \ \cdots \ t'r^{n-1}.$$

The two areas are then to each other in the ratio

$$s'\frac{r^m - 1}{r - 1} : t'\frac{r^n - 1}{r - 1}.$$

But $$s'r^m = s \text{ and } t'r^n = t,$$

so that this is the same as the ratio

$$s - s' : t - t'.$$

When l and k are incommensurable, we proceed as in other similar demonstrations.

This theorem is analogous to the one which we have proved about polygons : the area is proportional to the amount of rotation in excess of four right angles in going around the figure, for the rate of rotation in going along a boundary-curve is constant.

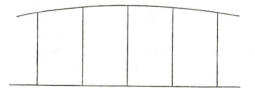

The locus of points at a given distance from a straight line is a curve which may be called an equidistant-curve. The perpendiculars from the different points of this curve upon the base line are equal and may be called axes of the curve.

An equidistant-curve fits upon itself when revolved through two right angles about one of its axes or when slid along upon itself. It has a constant curvature.

It can be proved, exactly as in the case of two boundary-curves having the same set of axes, that arcs on an equidistant-curve are proportional to the segments cut off by the axes at their extremities on the base line or on any other equidistant-curve having the same set of axes.

4. Theorem. *The boundary-curve is a limiting curve between the circle and the equidistant-curve ; it may be regarded as a circle with infinitely large radius, or as an equidistant-curve whose base line is infinitely distant.*

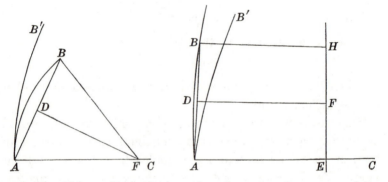

Proof. Take a line of given length, $AB = 2\,a$ say, making an angle, A, with a fixed line, AC. Construct another angle at B equal to the angle A, and draw a perpendicular to AB at its middle point, D.

If the angle at A is sufficiently small, we have an isosceles triangle with AB for base, and its vertex at a point, F, on AC. With F as centre, we can draw a circle through the points A and B. Now let the angle at A gradually increase, the rest of the figure varying so as to keep the construction. F will move off indefinitely, and when $A = \Pi\,(a)$ the three lines AF, BF, and DF will become parallel, and B will become a point on the boundary-curve AB', which has AC for axis.

On the other hand, if the angle at A were taken acute, but greater than $\Pi\,(a)$, we should have three lines, AE, BH, and

DF, perpendicular to a line, *EH*, the base line of an equidistant-curve through the points *A* and *B*.   Now let the angle *A* gradually decrease, the rest of the figure varying so as to preserve the construction.   The quadrilateral *ADFE*, having three right angles and the fourth angle *A* decreasing, must increase in area.   We get this same movement if we think of *AD* and *DF* remaining fixed in the plane while *AE* revolves about *A*, making the angle *A* decrease.   Thus the only way in which the area of the quadrilateral can increase is for *EH* to move off along on *AC* and become more and more remote from *A*. When *A* becomes equal to $\Pi(a)$, *BH* and *DF* become parallel to *AC*, and *B* falls on the boundary-curve *AB'*.

Calling the radius of a circle axis, we find that circles, boundary-curves, and equidistant-curves have many properties in common:

The perpendicular erected at the middle point of any chord is an axis.   In particular, a tangent is perpendicular to the axis drawn from its point of contact.   These are curves cutting at right angles a system of lines through a point, a system of parallel lines, and the perpendiculars to a given line, respectively.

Two of these curves having the same set of axes cut off equal lengths on all these axes, and the ratio of corresponding arcs on two such curves is a constant depending only on the way in which they divide the axes.

Three points determine one of these curves; that is, through any three points not in a straight line we can draw a curve which shall be either a circle, a boundary-curve, or an equidistant-curve, and through any three points only one such curve can be drawn.   Any triangle may be inscribed in one and only one of these curves.

Each of these curves can be moved on itself or revolved about any axis through 180° into coincidence with itself.

A boundary-surface or orisphere is a surface generated by the revolution of a boundary-curve about one of its axes.

5. Theorem. *Any line parallel to the axis of a boundary-surface may be regarded as axis.*

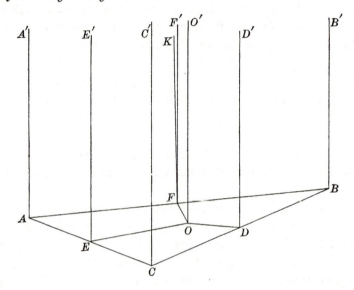

Let AA' be the axis, meeting the surface at A, and BB' a line parallel to the axis through any other point, B, of the surface; to prove that BB' may be regarded as axis.

Proof. Let C be a third point on the surface. Draw CC' through C, and through D, E, and F, the middle points of the sides of the plane triangle ABC, draw DD', EE', and FF' all parallel to AA'. Finally, let OO' be parallel to these lines and perpendicular to the plane ABC. The projecting planes of the other parallels all pass through OO' (see I, 9).

Since AA' is axis to the surface, EE' and FF' are perpendicular to AC and AB, respectively. Draw FK perpendicular to the plane ABC at F. It will lie in the projecting plane OFF'. AB, being perpendicular to FF' and to FK, is perpendicular to this plane, OFF', and therefore to OF. In the same way we prove that AC is perpendicular to OE. Therefore, BC is perpendicular to OD (Chap. I, I, 5). But OD is the intersection of the plane ABC with the plane ODD'. Hence, BC is perpendicular to this plane and to DD' (Chap. I, II, 15).

DD' being parallel to BB' lies in the plane determined by BB' and BC, and in this plane only one perpendicular can be drawn to BC at its middle point. Therefore, if we pass any plane through BB' and from B draw a chord to any other point, C, of its intersection with the surface, the perpendicular in this plane to BC, erected at the middle point of BC, will be parallel to BB'. This proves that the section is a boundary-curve, having BB' for axis, and that the surface can be generated by the revolution of such a boundary-curve around BB'.

Therefore, BB' may be regarded as axis of the surface.

A plane passed through an axis of a boundary-surface is called a principal plane. Every principal plane cuts the surface in a boundary-curve. Any other plane cuts the surface in a circle; for the surface may be regarded as a surface of revolution having for axis of revolution that axis which is perpendicular to the plane. This perpendicular may be called the axis of the circle, and the point where it meets the surface, the pole of the circle. The pole of a circle on a boundary-surface is at the same distance from all the points of the circle, distance being measured along boundary-lines on the surface.

Any two boundary-surfaces can be made to coincide, and a boundary-surface can be moved upon itself, any point to the position of any other point, and any boundary-curve through the first point to the position of any boundary-curve through the second point. We may say that a boundary-surface has a constant curvature, the same for all these surfaces. Figures on a boundary-surface can be moved about or put upon any other boundary-surface without altering their shape or size.

We can develop a Geometry on the boundary-surface. By *line* we mean the boundary-curve in which the surface is cut by a principal plane. The angle between two lines is the same as the diedral angle between the two principal planes which cut out the lines on the surface.

6. Theorem. *Geometry on the boundary-surface is the same as the ordinary Euclidean Plane Geometry.*

Proof. On two boundary-surfaces with the same system of parallel lines for axes corresponding triangles are similar; that is, corresponding angles are equal, having the same measures as the diedral angles which cut them out, and corresponding lines are proportional by (2). But we can place these figures on the same surface; therefore, on one boundary-surface we can have similar triangles. Thus, we can diminish the sides of a triangle without altering their ratios or the angles. We can do this indefinitely; for the ratio of corresponding lines on the two surfaces, being expressed by the function e^{ax} of the distance between them, can be made as large as we please by taking x sufficiently large. If we assume that figures on the boundary-surface become more and more like plane figures when we diminish indefinitely their size, it follows that a triangle on this surface approaches more and more the form of an infinitesimal plane triangle, for which the sum of the angles is two right angles, and the angles and sides have the same relations as in the Euclidean Plane Geometry. All the formulæ of Plane Trigonometry with which we are familiar hold, then, for triangles on the boundary-surface.

On the boundary-surface we have the "hypothesis of the right angle." Rectangles can be formed, and the area of a rectangle is proportional to the product of its base and altitude, while the area of a triangle is half of the area of a rectangle having the same base and altitude.

An equidistant-surface is a surface generated by the revolution of an equidistant-curve about one of its axes. It is the locus of points at a given perpendicular distance from a plane. Any perpendicular to the plane may be regarded as an axis, and the surface is a surface cutting at right angles a system of lines perpendicular to the plane. The surface has a constant curvature, fitting upon itself in any position.

III. TRIGONOMETRICAL FORMULÆ

1. Let ABC be a plane right triangle. Erect AA' perpendicular to its plane and draw BB' and CC' parallel to AA'. Draw a boundary-surface through A, having these lines for axes and forming the boundary-surface triangle $AB''C''$. Also construct the spherical triangle about the point B.

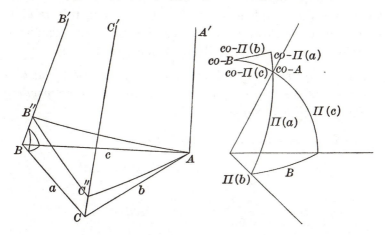

The angle A is the same in the plane triangle and in the boundary-surface triangle. The planes through AA' are perpendicular to ABC. Hence, the spherical triangle has a right angle at the vertex which lies on c, and BC being perpendicular to CA is perpendicular to the plane of CC' and AA'. Therefore, the plane BCC' is perpendicular to the plane ACC', and the diedral whose edge is BC has for plane angle the angle $ACC' = \Pi(b)$. Since the boundary-surface triangle is right-angled at C'', the angle B'', or what is the same thing, the diedral whose edge is BB', is the complement of the angle A.

In the spherical triangle the side opposite the right angle is $\Pi(a)$, the two sides about the right angle are $\Pi(c)$ and B, and the opposite angles are $\Pi(b)$ and $90° - A$.

Applying to these quantities the trigonometrical formulæ for spherical right triangles, we get at once the relations that connect the sides and angles of plane right triangles.

Produce to quadrants the two sides about the angle whose value is the complement of A. We form in this way a spherical right triangle in which the side opposite the right angle is the complement of $\Pi(c)$, the two sides about the right angle are the complements of $\Pi(a)$ and $\Pi(b)$, and their opposite angles are the complements of B and A. From this triangle we deduce the following rule for passing from the formulæ of spherical right triangles to those of plane triangles:

Interchange the two angles (or the two sides) and everywhere use the complementary function, taking the corresponding angle of parallelism for the sides.

The formulæ for spherical right triangles are

$$\sin A = \frac{\sin a}{\sin c}. \qquad\qquad \sin B = \frac{\sin b}{\sin c}.$$

$$\cos A = \frac{\tan b}{\tan c}. \qquad\qquad \cos B = \frac{\tan a}{\tan c}.$$

$$\tan A = \frac{\tan a}{\sin b}. \qquad\qquad \tan B = \frac{\tan b}{\sin a}.$$

$$\sin A = \frac{\cos B}{\cos b}. \qquad\qquad \sin B = \frac{\cos A}{\cos a}.$$

$$\cos c = \cos a \cos b.$$

$$\cos c = \cot A \cot B.$$

From these, by the rule given on the previous page, we derive the following formulæ for plane right triangles:

$$\cos B = \frac{\cos \Pi\,(a)}{\cos \Pi\,(c)}. \qquad\qquad \cos A = \frac{\cos \Pi\,(b)}{\cos \Pi\,(c)}.$$

$$\sin B = \frac{\cot \Pi\,(b)}{\cot \Pi\,(c)}. \qquad\qquad \sin A = \frac{\cot \Pi\,(a)}{\cot \Pi\,(c)}.$$

$$\cot B = \frac{\cot \Pi\,(a)}{\cos \Pi\,(b)}. \qquad\qquad \cot A = \frac{\cot \Pi\,(b)}{\cos \Pi\,(a)}.$$

$$\cos B = \frac{\sin A}{\sin \Pi\,(b)}. \qquad\qquad \cos A = \frac{\sin B}{\sin \Pi\,(a)}.$$

$$\sin \Pi\,(c) = \sin \Pi\,(a)\,\sin \Pi\,(b).$$

$$\sin \Pi\,(c) = \tan A \tan B.^{*}$$

We can obtain the formulæ for oblique plane triangles by dropping a perpendicular from one vertex upon the opposite side, thus forming two right triangles.

2. Take the relation

$$\sin \Pi\,(a) = \frac{\sin B}{\cos A}.$$

Let p, q, and r be the sides of the triangle $AB''C''$ of our last demonstration and p', q', and r the corresponding sides

---

* We can arrange the parts of a right triangle so as to apply Napier's rules; namely, the arrangement would be

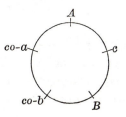

of the triangle formed in the same way on a boundary-surface tangent to the plane ABC at B.

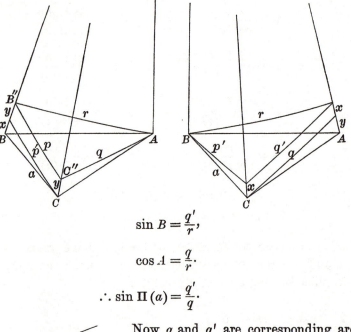

$$\sin B = \frac{q'}{r},$$

$$\cos A = \frac{q}{r}.$$

$$\therefore \sin \Pi(a) = \frac{q'}{q}.$$

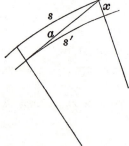

Now q and q' are corresponding arcs on two boundary-curves which have the same set of parallel lines as axes, and their distance apart, x, is the distance from a boundary-curve of the extremity of a tangent of arbitrary length, a. Thus, we have for corresponding arcs

$$\frac{s'}{s} = \sin \Pi(a).$$

3. To MN, a given straight line, erect a perpendicular at a point, O, and on this perpendicular lay off $OA = y$ below MN, and OB and BP each equal to x above MN, x and y being any arbitrary lengths. At P draw PR perpendicular to OP and extending towards the left, and through B draw EF making with OP an angle $\Pi(x)$, and therefore parallel on one side to ON and on the other side to PR. Finally, draw AK and AH, the two parallels to EF through A.

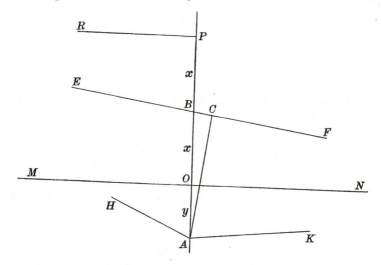

At the point A we have four angles of parallelism:

$$CAK = CAH = \Pi(AC),$$
$$OAK = \Pi(y),$$
$$PAH = \Pi(y + 2x).$$

Therefore, $\qquad \Pi(y) = \Pi(AC) + BAC,$

and $\qquad \Pi(y + 2x) = \Pi(AC) - BAC.$

Now in the right triangle ABC

$$\cos \Pi(y + x) = \frac{\cos \Pi(AC)}{\cos BAC},$$

or $\qquad \dfrac{1 - \cos \Pi(y + x)}{1 + \cos \Pi(y + x)} = \dfrac{\cos BAC - \cos \Pi(AC)}{\cos BAC + \cos \Pi(AC)}$

$$= \frac{\sin \frac{1}{2}[\Pi(AC) + BAC]\sin \frac{1}{2}[\Pi(AC) - BAC]}{\cos \frac{1}{2}[\Pi(AC) + BAC]\cos \frac{1}{2}[\Pi(AC) - BAC]};$$

whence,

$$\tan^2 \tfrac{1}{2} \Pi (y + x) = \tan \tfrac{1}{2} \Pi (y) \tan \tfrac{1}{2} \Pi (y + 2x).$$

$\tan \tfrac{1}{2} \Pi (x)$ is then a function of x, say $f(x)$, satisfying the condition

$$[f(y + x)]^2 = f(y) f(y + 2x),$$

or

$$\frac{f(y + x)}{f(y)} = \frac{f(y + 2x)}{f(y + x)},$$

and putting successively in this equation $y + x$, $y + 2x$, etc., for y, we may add

$$= \frac{f(y + 3x)}{f(y + 2x)} = \cdots = \frac{f(y + nx)}{f[y + (n - 1)x]}.$$

$\Pi (0) = \dfrac{\pi}{2}$ and $\tan \tfrac{1}{2} \Pi (0) = 1$; therefore, putting $y = 0$ in the first and last of all these fractions, we have

$$f(x) = \frac{f(nx)}{f[(n - 1)x]},$$

or

$$f(nx) = f[(n - 1)x] f(x).$$

$$\therefore f(nx) = [f(x)]^n.$$

This equation is characteristic of the exponential function.* $\Pi (x)$ being an acute angle, $\tan \tfrac{1}{2} \Pi (x) < 1$; therefore, we may write $f(1) = e^{-a'}$, so that $f(x) = e^{-a'x}$. a' depends on the unit of measure; we will take the unit so that $a' = 1$. Finally, since $\Pi (- x) = \pi - \Pi (x)$,

$$\tan \tfrac{1}{2} \Pi (- x) = \cot \tfrac{1}{2} \Pi (x) = [\tan \tfrac{1}{2} \Pi (x)]^{-1}.$$

That is, for all real values of x

$$\tan \tfrac{1}{2} \Pi (x) = e^{-x},$$

* See footnote, p. 45.

or
$$\frac{1 - \cos \Pi(x)}{\sin \Pi(x)} = \cos ix + i \sin ix.*$$

* i stands for $\sqrt{-1}$. The best way to get the relations between the exponential and trigonometrical functions is by their developments in series:

$$e^x = 1 + x + \frac{x^2}{\underline{2}} + \cdots + \frac{x^n}{\underline{n}} + \cdots,$$

$$\cos x = 1 - \frac{x^2}{\underline{2}} + \frac{x^4}{\underline{4}} - \cdots + (-1)^n \frac{x^{2n}}{\underline{2n}} + \cdots,$$

$$\sin x = x - \frac{x^3}{\underline{3}} + \frac{x^5}{\underline{5}} - \cdots + (-1)^n \frac{x^{2n+1}}{\underline{2n+1}} + \cdots.$$

These series are convergent for all values of x.

Putting ix for x, we have

$$e^{ix} = 1 + ix - \frac{x^2}{\underline{2}} - \frac{ix^3}{\underline{3}} + \cdots$$

$$= 1 - \frac{x^2}{\underline{2}} + \frac{x^4}{\underline{4}} - \cdots + i\left(x - \frac{x^3}{\underline{3}} + \frac{x^5}{\underline{5}} - \cdots\right);$$

i.e.,
$$e^{ix} = \cos x + i \sin x.$$

Also
$$e^{-ix} = \cos x - i \sin x.$$

$$\therefore \cos x = \tfrac{1}{2}(e^{ix} + e^{-ix}),$$

$$\sin x = \frac{1}{2i}(e^{ix} - e^{-ix}).$$

Again, putting ix for x, we have

$$e^x = \cos ix - i \sin ix,$$

$$e^{-x} = \cos ix + i \sin ix ;$$

and
$$\cos ix = \tfrac{1}{2}(e^x + e^{-x}),$$

$$\sin ix = -\frac{1}{2i}(e^x - e^{-x}).$$

$$\cos ix = 1 + \frac{x^2}{\underline{2}} + \frac{x^4}{\underline{4}} + \cdots,$$

$$\sin ix = ix\left(1 + \frac{x^2}{\underline{3}} + \frac{x^4}{\underline{5}} + \cdots\right).$$

For real values of x, $\cos ix$ and $\dfrac{\sin ix}{ix}$ are real and positive, and vary from 1 to ∞ as x varies from 0 to ∞.

In the equation $\cos^2 ix + \sin^2 ix = 1$, the first term is real and positive for real values of x, the second term is real and negative ; therefore, $\sin ix$ is in absolute value less than $\cos ix$, and $\tan ix$ is in absolute value less than 1. $\tan ix$ varies in absolute value from 0 to 1 as x varies from 0 to ∞.

Changing the sign of x, we have

$$\frac{1 + \cos \Pi(x)}{\sin \Pi(x)} = \cos ix - i \sin ix,$$

and, adding and subtracting,

$$\frac{1}{\sin \Pi(x)} = \cos ix,$$

$$\cot \Pi(x) = -i \sin ix.$$

The nature of the angle of parallelism is, therefore, expressed by the equations

$$\sin \Pi(x) = \frac{1}{\cos ix},$$

$$\tan \Pi(x) = \frac{i}{\sin ix},$$

$$\cos \Pi(x) = \frac{\tan ix}{i}.$$

4. Substituting in the formulæ of plane right triangles, we find that they reduce to those of spherical right triangles with ia, ib, and ic for a, b, and c, respectively. The formulæ of oblique triangles are obtained from those of right triangles in the same way as on the sphere, and thus all the formulæ of Plane Trigonometry are obtained from those of Spherical Trigonometry simply by making this change.

As fundamental formulæ for oblique triangles we write

$$\frac{\sin A}{\sin ia} = \frac{\sin B}{\sin ib} = \frac{\sin C}{\sin ic},$$

$$\cos ia = \quad \cos ib \cos ic + \sin ib \sin ic \cos A,$$

$$\cos A = -\cos B \cos C + \sin B \sin C \cos ia.$$

In the notation of the Π-function, these are

$$\sin A \tan \Pi(a) = \sin B \tan \Pi(b) = \sin C \tan \Pi(c),$$

$$\frac{\sin \Pi (b) \sin \Pi (c)}{\sin \Pi (a)} = 1 - \cos \Pi (b) \cos \Pi (c) \cos A,$$

$$\cos A = -\cos B \cos C + \frac{\sin B \sin C}{\sin \Pi (a)}.$$

5. Since for very small values of x we have approximately

$$\sin ix = ix,$$

$$\cos ix = 1 + \frac{x^2}{2},$$

$$\tan ix = ix,$$

our formulæ for infinitesimal triangles reduce to

$$\frac{\sin A}{a} = \frac{\sin B}{b} = \frac{\sin C}{c},$$

$$a^2 = b^2 + c^2 - 2\,bc \cos A,$$

$$\cos A = -\cos (B + C).$$

6. Triangles on an equidistant-surface are similar to their projections on the base plane; that is, they have the same angles and their sides are proportional. Thus the formulæ of Plane Trigonometry hold for any equidistant-surface if with the letters representing the sides we put, besides i, a constant factor depending on the distance of the surface from the plane.

CHAPTER III

THE ELLIPTIC GEOMETRY

In the hypothesis of the obtuse angle a straight line is of finite length and returns into itself. This length is the same for all lines, since any two lines can be made to coincide. Two straight lines always intersect, and two lines perpendicular to a third intersect at a point whose distance from the third on either line is half the entire length of a straight line.

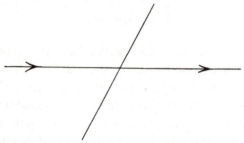

1. A straight line does not divide the plane. Starting from the point of intersection of two lines and passing along one of them a certain finite distance, we come to the intersection point again without having crossed the other line. Thus, we can pass from one side of the line to the other without having crossed it.

There is one point through which pass all the perpendiculars to a given line. It is called the pole of that line, and the line is its polar. Its distance from the line is half the entire length of a straight line, and the line is the locus of points at this distance from its pole. Therefore, if the pole of one line lies on another, the pole of the second lies on the first, and the intersection of two lines is the pole of the line joining their poles.

The locus of points at a given distance from a given line is a circle having its centre at the pole of the line. The straight line is a limiting form of a circle when the radius becomes equal to half the entire length of a line.

We can draw three lines, each perpendicular to the other two, forming a trirectangular triangle. It is also a self-polar triangle; each vertex is the pole of the opposite side.

2. All the perpendiculars to a plane in space meet at a point which is the pole of the plane. It is the centre of a system of spheres of which the plane is a limiting form when the radius becomes equal to half the entire length of a straight line.

Figures on a plane can be projected from similar figures on any sphere which has the pole of the plane for centre. That is, they have equal angles and corresponding sides in a constant ratio that depends only on the radius of the sphere. Two corresponding angles are equal, because they are the same as the diedral angles formed by the two planes through the centre of the sphere which cut the sphere and the plane in the sides of the angles. Corresponding lines are proportional; for if two arcs on the sphere are equal, their projections on the plane are equal; and that, in general, two arcs have the same ratio as their projections on the plane is proved, first when they are commensurable, and by the method of limits when they are incommensurable.

Geometry on a plane is, therefore, like Spherical Geometry, but the plane corresponds to only half a sphere, just as the diameters of a sphere correspond to the points of half the surface. Indeed, the points and straight lines of a plane correspond exactly to the lines and planes through a point, but we can realize the correspondence better that compares the plane with the surface of a sphere. If we can imagine that the points on the boundary of a hemisphere at opposite extremities of diameters are coincident, the hemisphere will correspond to the elliptic plane. There is no particular line of the plane that plays the part of boundary. All lines of the plane are alike; the plane is unbounded, but not infinite in extent.

The entire straight line corresponds to a semicircle. We will take such a unit for measuring length that the entire length of a line shall be π; the formulæ of Spherical Trigonometry will then apply without change to our plane. Distances on a line will then have the same measure as the angles which they subtend at the pole of the line, and the angle between two lines will be equal to the distance between their poles. The distance from any point to its polar, half the entire length of a straight line, may then be called a quadrant.

We can form a self-polar tetraëdron by taking three mutually perpendicular planes and the plane which has their intersection for pole. The vertices of this tetraëdron are the poles of the opposite faces. At each vertex is a trirectangular triedral, and each face is a trirectangular triangle.

3. Theorem. *All the planes perpendicular to a fixed line intersect in another fixed line, called its polar or conjugate. The relation is reciprocal, and all the points of either line are at a quadrant's distance from all the points of the other.*

Proof. Let the two planes perpendicular to the line AB at H and K intersect in CD. Pass a plane through AB and R, any point of CD. This plane will intersect the two given planes in HR and KR. HR and KR are perpendicular to AB; therefore, R is at a quadrant's distance from H and K. R is then the pole of AB in the plane determined by AB and R, and is at a quadrant's distance from every point of AB. But R is any point of CD; therefore, any point of either line is at a quadrant's distance from each point of the other line, and a point which is at a quadrant's distance from one line lies in

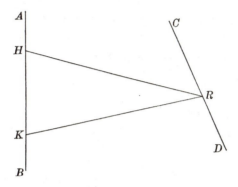

the other line. Again, any point, H, of AB, being at a quadrant's distance from all the points of CD, is the pole of CD in the plane determined by it and CD. Thus, HR and KR are both perpendicular to CD, and the plane determined by AB and R is perpendicular to CD.

The opposite edges of a self-polar tetraëdron are polar lines.

All the lines which intersect a given line at right angles intersect its polar at right angles. Therefore, the distances of any point from two polar lines are measured on the same straight line and are together equal to a quadrant. Two points which are equidistant from one line are equidistant from its polar.

The locus of points which are at a given distance from a fixed line is a surface of revolution having both this line and its polar as axes. We may call it a surface of double revolution. The parallel circles about one axis are meridian curves for the other axis. If a solid body, or, we may say, all space, move along a straight line without rotating about it, it will rotate about the conjugate line as an axis without sliding

along it. A motion along a straight line combined with a
rotation about it is called a screw motion. A screw motion
may then be described as a rotation about each of two conjugate lines or as a sliding along each of two conjugate lines.

4. Theorem. *In the elliptic geometry there are lines not
in the same plane which have an infinite number of common
perpendiculars and are everywhere equidistant.*

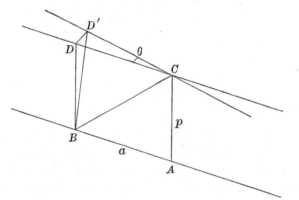

Given any two lines in the same plane and their common perpendicular. If we go out on these lines in either direction from
the perpendicular, they approach each other. Now revolve
one of them about this perpendicular so that they are no longer
in the same plane. After a certain amount of rotation the lines
will have an infinite number of common perpendiculars and be
equidistant throughout their entire length.

Proof. Let p be the length of the common perpendicular
AC, and take points B and D on the two lines on the same
side of this perpendicular at a distance, a.

$BD < p$, but if CD revolve about AC, BD will become longer
than p by the time CD is revolved through a right angle; for
BCD will then be a right triangle, with BD for hypothenuse
and BC, the hypothenuse of the triangle ABC, for one of
its sides, so that we shall have $BD > BC$ and $BC > AC$.

Suppose, when CD has revolved through an angle, θ, BD becomes equal to p and takes the position BD'. The triangles ABC and $D'BC$ are equal, having corresponding sides equal. Therefore, BD' is perpendicular to CD'. BD' is also perpendicular to BA; for if we take the diedral A-BC-D' and place it upon itself so that the positions of B and C shall be interchanged, A will fall on the position of D', and D' on the position of A, and the angle $D'BA$ must equal the angle ACD'. Therefore, BD' as well as CA is a common perpendicular to the lines AB and CD'.

Now at the point C we have a triedral whose three edges are CB, CD, and CD'. Moreover, the diedral along the edge CD is a right diedral; therefore, the three face angles of the triedral satisfy the same relations as do the three sides of a spherical right triangle; namely,

$$\cos BCD' = \cos BCD \cos DCD'.$$

But $\qquad BCD = \dfrac{\pi}{2} - ACB \quad$ and $\quad BCD' = ABC.$

Hence, this relation may be written

$$\cos ABC = \sin ACB \cos \theta.$$

Again, in the right triangle ABC

$$\sin ACB = \frac{\cos ABC}{\cos p}.$$

$$\therefore \cos \theta = \cos p,$$

or, since θ and p are less than $\dfrac{\pi}{2}$,

$$\theta = p.$$

The angle θ, therefore, does not depend upon a. If we take any two lines in a plane and turn one about their common perpendicular through an angle equal in measure to the length of that perpendicular, the two lines will then be everywhere equidistant.

As we have no parallel lines in the ordinary sense in this Geometry, the name *parallel* has been applied to lines of this kind.   They have many properties of the parallel lines of Euclidean Geometry.

Through any point two lines can be drawn parallel to a given line.   These are of two kinds, sometimes distinguished as right-wound and left-wound.   They lie entirely on a surface of double revolution, having the given line as axis.   The surface is, therefore, a ruled surface and has on it two sets of rectilinear generators like the hyperboloid of one sheet.

CHAPTER IV

ANALYTIC NON–EUCLIDEAN GEOMETRY

We shall use the ordinary polar coördinates, ρ and θ, and for the rectangular coördinates, x and y, of a point, we shall use the intercepts on the axes made by perpendiculars through the point to the axes. The formulæ depend upon the trigonometrical relations, and in our two Geometries differ only in the use of the imaginary factor i with lengths of lines.

I. HYPERBOLIC ANALYTIC GEOMETRY

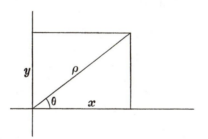

1. The relations between polar and rectangular coördinates :

The angles at the origin which the radius vector makes with the axes are complementary. From the two right triangles we have

$$\tan ix = \cos \theta \tan i\rho,$$

$$\tan iy = \sin \theta \tan i\rho.$$

Therefore,

$$\tan^2 i\rho = \tan^2 ix + \tan^2 iy,$$

$$\tan \theta = \frac{\tan iy}{\tan ix}.$$

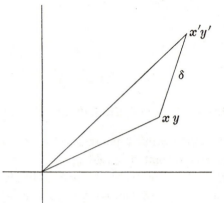

2. The distance, δ, between two points :

$$\cos i\delta = \cos i\rho \cos i\rho' + \sin i\rho \sin i\rho' \cos (\theta' - \theta).$$

δ and one of the points being fixed, this may be regarded as the polar equation of a circle.

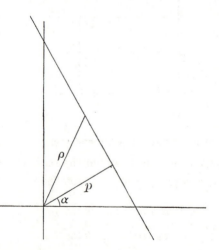

3. The equation of a line :

Let p be the length of the perpendicular from the origin upon the line, and α the angle which the perpendicular makes with the axis of x. From the right triangle formed with this perpendicular and ρ we have

$$\tan i\rho \cos (\theta - \alpha) = \tan ip.$$

This is the polar equation of the line. We get the equation in x and y by expanding and substituting; namely,

$$\cos \alpha \tan ix + \sin \alpha \tan iy = \tan ip.$$

The equation $\quad a \tan ix + b \tan iy = i$

represents a line for which

$$a^2 + b^2 = \frac{-1}{\tan^2 ip}.$$

Now, for real values of p, $- \tan^2 ip < 1$ (see footnote, p. 59). The line is therefore real if a and b are real, and if

$$a^2 + b^2 > 1.$$

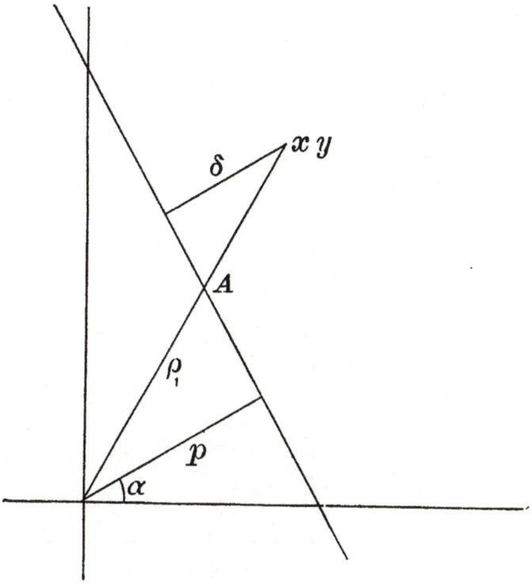

4. **The distance, δ, of a point from a line :**

Let the radius vector to the point intersect the line at A, and let ρ_1 be the radius vector to A. We have two right triangles with equal angles at A, and from the expressions for the sines of these angles we get the equation

$$\frac{\sin i\delta}{\sin i(\rho - \rho_1)} = \frac{\sin ip}{\sin i\rho_1}.$$

This equation holds for all points, $x\,y$, of the plane, δ being negative when the point is on the same side of the line as the origin, and zero when the point is on the line.

$$\sin i\delta = \frac{\sin ip}{\tan i\rho_1} \sin i\rho - \sin ip \cos i\rho.$$

Now, $\tan i\rho_1 = \dfrac{\tan ip}{\cos(\theta - \alpha)}.$

$$\therefore \frac{\sin ip}{\tan i\rho_1} \sin i\rho = \sin i\rho \cos ip \cos(\theta - \alpha),$$

and $\sin i\delta = \cos i\rho \cos ip \left[\tan i\rho \cos(\theta - \alpha) - \tan ip\right].$

δ being fixed, this may be regarded as the polar equation of an equidistant-curve.

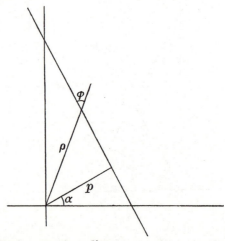

5. The angle between two lines :

ϕ being the angle which a line makes with the radius vector at any point, we have

$$\cos \phi = \cos ip \sin (\theta - \alpha),$$

$$\sin \phi = \frac{\sin ip}{\sin i\rho}.$$

For two lines intersecting at this point,

$$\sin \phi_1 \sin \phi_2 = \frac{\sin ip_1 \sin ip_2}{\sin^2 i\rho}$$

$$= \sin ip_1 \sin ip_2 + \frac{\sin ip_1 \sin ip_2}{\tan^2 i\rho}.$$

Now, from the equation of the line

$$\frac{\sin ip_1}{\tan i\rho} = \cos ip_1 \cos (\theta - \alpha_1),$$

$$\frac{\sin ip_2}{\tan i\rho} = \cos ip_2 \cos (\theta - \alpha_2);$$

so that $\quad \sin \phi_1 \sin \phi_2 = \sin ip_1 \sin ip_2$

$$+ \cos ip_1 \cos ip_2 \cos (\theta - \alpha_1) \cos (\theta - \alpha_2).$$

Again, $\quad \cos \phi_1 \cos \phi_2 = \cos ip_1 \cos ip_2 \sin (\theta - \alpha_1) \sin (\theta - \alpha_2).$

Adding these equations, we have

$$\cos (\phi_2 - \phi_1) = \sin ip_1 \sin ip_2 + \cos ip_1 \cos ip_2 \cos (\alpha_2 - \alpha_1).$$

Two lines are perpendicular if

$$\cos (\alpha_2 - \alpha_1) + \tan ip_1 \tan ip_2 = 0.$$

The lines $\quad a \tan ix + b \tan iy = i,$

$$a' \tan ix + b' \tan iy = i$$

are perpendicular if $\quad aa' + bb' = 1.$

6. The equation of a circle in x and y:

$$\sin i\rho \cos \theta = \cos i\rho \tan ix,$$

$$\sin i\rho \sin \theta = \cos i\rho \tan iy;$$

also, $\quad \cos i\rho = \dfrac{1}{\sqrt{1 + \tan^2 i\rho}} = \dfrac{1}{\sqrt{1 + \tan^2 ix + \tan^2 iy}}.$

The equation of a circle may, therefore, be written

$$(1 + \tan^2 ix + \tan^2 iy)(1 + \tan^2 ix' + \tan^2 iy')\cos^2 i\delta$$
$$= (1 + \tan ix \tan ix' + \tan iy \tan iy')^2.$$

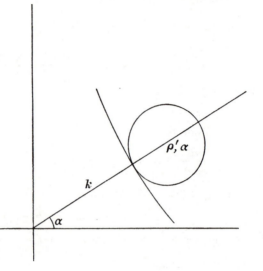

7. The equation of a boundary-curve:

Let the axis of the boundary-curve which passes through the origin make an angle, α, with the axis of x, and let the point where the boundary-curve cuts this axis be at a distance, k, from the origin, positive if the origin is on the convex side of the curve, negative if the origin is on the concave side of the curve. The boundary-curve is the limiting position of a circle whose centre, on this axis, moves off indefinitely.

ρ' being the radius vector to the centre, the radius of the circle is $\rho' - k$, and its equation may be written

$$\cos i(\rho' - k) = \cos i\rho \cos i\rho' + \sin i\rho \sin i\rho' \cos(\theta - \alpha),$$

or, expanding and dividing by $\cos i\rho'$,

$$\cos ik + \tan i\rho' \sin ik = \cos i\rho + \sin i\rho \tan i\rho' \cos(\theta - \alpha).$$

Now, let ρ' increase indefinitely. $\tan i\rho'$ tends to the limit i, so that the limit of the first member of the equation is

$$\cos ik + i \sin ik, \quad \text{or} \quad e^{-k},$$

and the polar equation of the curve is

$$e^{-k} = \cos i\rho \,[1 + i \tan i\rho \cos (\theta - \alpha)];$$

or, in $x\,y$ coördinates,

$$(1 + \tan^2 ix + \tan^2 iy)\, e^{-2k}$$
$$= (1 + i \cos \alpha \tan ix + i \sin \alpha \tan iy)^2.$$

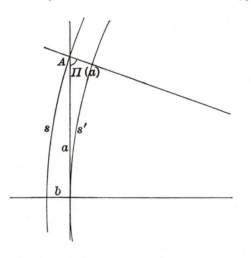

Let k be negative and equal, say, to $-b$, and let $\alpha = 0$; also, let a be the ordinate of the point A where the curve cuts the axis of y.

Substituting in the equation, we find

$$e^b = \cos ia.$$

Through A draw a line parallel to the axis of x, and, therefore, making an angle, $\Pi\,(a)$, with the axis of y. If we draw a boundary-curve through the origin having the same set of parallel lines for axes, so that the two boundary-curves cut off a distance, b, on these axes, we know that the ratio of corresponding arcs is

$$\frac{s'}{s} = \sin \Pi (a) = \frac{1 \cdot}{\cos ia}; \qquad \text{(See p. 56.)}$$

therefore, $\qquad\qquad \frac{s'}{s} = e^{-b}. \qquad\qquad$ (See p. 45.)

8. The equation of an equidistant-curve :

The polar equation of (4) reduced to an equation in x and y takes the form

$$(1 + \tan^2 ix + \tan^2 iy) \sin^2 i\delta$$
$$= \cos^2 ip \,(\cos \alpha \tan ix + \sin \alpha \tan iy - \tan ip)^2.$$

9. Comparison of the three equations :

The equation

$$(1 + \tan^2 ix + \tan^2 iy)\, c^2 = -\,(i - a \tan ix - b \tan iy)^2$$

represents a circle, a boundary-curve, or an equidistant-curve, according as $a^2 + b^2 < 1, \; = 1, \; > 1$, respectively.

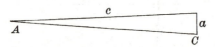

10. Differential formulæ :

Suppose we have an isosceles triangle in which the angle A at the vertex diminishes indefinitely.   In the formula

$$\frac{\sin A}{\sin ia} = \frac{\sin C}{\sin ic}$$

we may put for $\qquad \sin A, \quad \sin ia, \quad \sin C$;
$$A, \qquad ia, \qquad 1,$$

respectively.   Therefore,

(I.) $\qquad\qquad\qquad ia = \sin ic \cdot A.$

Corollary. *In a circle of radius* **r**, *the ratio of any arc to the angle subtended at the centre is sin i***r**.

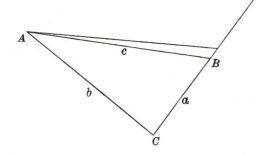

Again, in the right triangle ABC, let the hypothenuse c revolve about the vertex A. Differentiating the equation

$$\sin A = \frac{\cos B}{\cos ib},$$

where b is constant, we have

$$\cos A\, dA = -\frac{\sin B\, dB}{\cos ib}.$$

But
$$\sin B = \frac{\cos A}{\cos ia};$$

$$\therefore\ dB = -\cos ia \cos ib\, dA,$$

or (II.) $\qquad\qquad dB = -\cos ic\, dA.$

Now, using polar coördinates, we have an infinitesimal right triangle whose hypothenuse, ds, makes an angle, say ϕ, with the radius vector (see figure on page 78). The two sides about the right angle are $d\rho$ and $\dfrac{\sin i\rho}{i}\, d\theta$;

therefore, $\qquad\qquad ds^2 = d\rho^2 - \sin^2 i\rho\, d\theta^2,$

$$\tan \phi = \frac{\sin i\rho}{i}\frac{d\theta}{d\rho}.$$

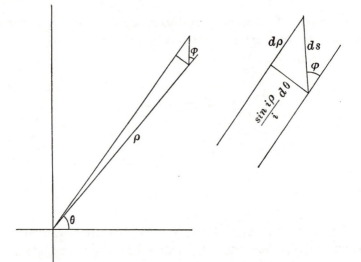

For two arcs cutting at right angles, let d' denote differentiation along the second arc:

$$\frac{\sin i\rho}{i} \frac{d\theta}{d\rho} = -\frac{i}{\sin i\rho} \frac{d'\rho}{d'\theta},$$

or

$$\frac{d\rho}{d\theta} \frac{d'\rho}{d'\theta} = \sin^2 i\rho.$$

11. Area:

It equals

$$\iint \frac{\sin i\rho}{i} \, d\rho \, d\theta.^*$$

We will consider only the case where the origin is within the area to be computed and where each radius vector meets the bounding curve once, and only once.

Integrating with respect to ρ, from $\rho = 0$, we have

$$\int_0^{2\pi} (\cos i\rho - 1) \, d\theta,$$

or

$$\int_0^{2\pi} \cos i\rho \, d\theta \ - 2\pi.$$

* The unit of area being so chosen that the area of an infinitesimal rectangle may be expressed as the product of its base and altitude.

Suppose P and P' are two "consecutive" points on the curve, PM and $P'M'$ the tangents at these points, and ϕ the

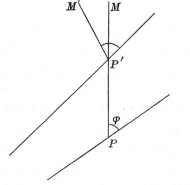

angle which the tangent makes with the radius vector. The angle $MP'M'$ indicates the amount of turning or rotation at these points as we go around the curve.

Now, by (II.),

$$MP'M' = d\phi + \cos i\rho \, d\theta.$$

In going around the curve, ϕ may vary but finally returns to its original value. That is, for our curve

$$\int d\phi = 0,$$

and the amount of rotation is

$$\int_0^{2\pi} \cos i\rho \, d\theta.$$

Hence, the area is equal to the excess over four right angles in the amount of rotation as we go around the curve. This theorem can be extended to any finite area.

12. A modified system of coördinates :

Our equations take simple forms if we write iu for $\tan ix$, iv for $\tan iy$, ir for $\tan i\rho$, and so on for all lengths of lines.

Thus, we have $\qquad u^2 + v^2 = r^2.$*

The equation of a line is

$$au + bv = 1,$$

and the equation

$$(1 - u^2 - v^2)\,c^2 = (1 - au - bv)^2$$

represents a circle, a boundary-curve, or an equidistant-curve, according as $a^2 + b^2 < 1,\ = 1,\ > 1$, respectively.

II. ELLIPTIC ANALYTIC GEOMETRY

The Elliptic Analytic Geometry may be developed just as we have developed the Hyperbolic Analytic Geometry, and the formulæ are the same with the omission of the factor i. But these formulæ are also very easily obtained from the relation of line and pole, and we shall produce them in this way.

The formulæ of Elliptic Plane Analytic Geometry may be applied to a sphere in any of our three Geometries.

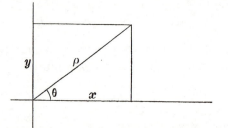

1. The relations between polar and rectangular coördinates :

$$\tan x = \cos \theta \tan \rho, \qquad \tan y = \sin \theta \tan \rho \,;$$

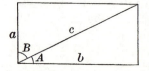

* If we draw a quadrilateral with three right angles and the diagonal to the acute angle, and use a, b, and c in the same way that u, v, and r are used above, the five parts lettered in the figure have the relations of a right triangle in the Euclidean Geometry ; *e.g.*,

$$a^2 + b^2 = c^2, \quad \sin A = \frac{a}{c}, \quad \text{etc.}$$

therefore, $\tan^2 \rho = \tan^2 x + \tan^2 y,$

$$\tan \theta = \frac{\tan y}{\tan x}.*$$

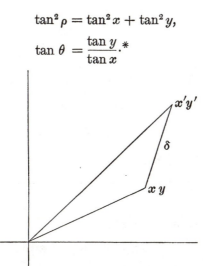

2. The distance, δ, between two points :

$$\cos \delta = \cos \rho \cos \rho' + \sin \rho \sin \rho' \cos (\theta' - \theta).$$

This may be regarded as the polar equation of a circle of radius δ, ρ' and θ' being the polar coördinates of the centre.

Now, $\sin \rho \cos \theta = \cos \rho \tan x,$

$\sin \rho \sin \theta = \cos \rho \tan y \,;$

also, $\cos \rho = \dfrac{1}{\sqrt{1 + \tan^2 \rho}} = \dfrac{1}{\sqrt{1 + \tan^2 x + \tan^2 y}}.$

The equation of a circle in rectangular coördinates may, therefore, be written

$$(1 + \tan^2 x + \tan^2 y)\,(1 + \tan^2 x' + \tan^2 y')\cos^2 \delta$$
$$= (1 + \tan x \tan x' + \tan y \tan y')^2.$$

---

* The line which has the origin for pole forms with the coördinate axes a trirectangular triangle, and x, y, and θ may be regarded as representing the directions of the given point from its three vertices.

On a sphere, if we take as origin the pole of the equator, ρ and θ are colatitude and longitude. x and y, one with its sign changed, are the "bearings" of the point from two points 90° apart on the equator.

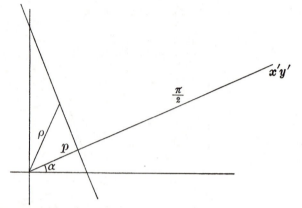

3. The equation of a line:

When $\delta = \dfrac{\pi}{2}$, the circle becomes a straight line. For this we have, therefore, the equation

$$\tan x \tan x' + \tan y \tan y' + 1 = 0.$$

$x'y'$ is the pole of the line.

From the equation

$$\tan \rho \cos (\theta - \alpha) = \tan p,$$

or

$$\cos \alpha \tan x + \sin \alpha \tan y = \tan p,$$

we find

$$\tan x' = -\frac{\cos \alpha}{\tan p},$$

$$\tan y' = -\frac{\sin \alpha}{\tan p}$$

as can be shown geometrically, the polar coördinates of this point being

$$p + \frac{\pi}{2}, \quad \alpha.$$

The equation $a \tan x + b \tan y + 1 = 0$

represents a real line for any real values of a and b.

4. The distance, δ, of a point from a straight line:

This is the complement of the distance between the point and the pole of the line; it is expressed by the equation

$$\sin \delta = - \cos \rho \sin p + \sin \rho \cos p \cos (\theta - \alpha)$$
$$= \cos \rho \cos p \, [\tan \rho \cos (\theta - \alpha) - \tan p].$$

5. The angle, φ, between two lines :

This is equal to the distance between their poles ; therefore,

$$\cos \phi = \sin p \sin p' + \cos p \cos p' \cos (\alpha' - \alpha).$$

The two lines $a \tan x + b \tan y + 1 = 0,$

$$a' \tan x + b' \tan y + 1 = 0$$

are perpendicular if $aa' + bb' + 1 = 0.$

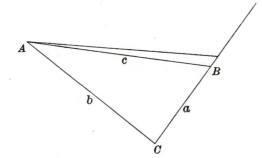

6. Differential formulæ :

The formula $$\frac{\sin A}{\sin a} = \frac{\sin C}{\sin c}$$

becomes, when A diminishes indefinitely,

(I.) $a = \sin c \cdot A.$

Corollary. *In a circle of radius* **r**, *the ratio of any arc to the angle subtended at the centre is sin* **r**.

From the right triangle ABC, if b remain fixed, we get, by differentiating the equation

$$\sin A = \frac{\cos B}{\cos b},$$

(II.) $$dB = -\cos c\, dA.$$

Thus, we have for differential formulæ in polar coördinates

$$ds^2 = d\rho^2 + \sin^2 \rho\, d\theta^2,$$

$$\tan \phi = \sin \rho\, \frac{d\theta}{d\rho}; *$$

* If ϕ is constant, as in the logarithmic spiral of Euclidean Geometry, we can integrate this equation ; namely,

$$\tan \phi\, \frac{d\rho}{\sin \rho} = d\theta.$$

$$\therefore\ \tan \phi \log \tan \frac{\rho}{2} = \theta + c,$$

or

$$\tan \frac{\rho}{2} = e^{\frac{\theta + c}{\tan \phi}}.$$

Writing c' for $e^{\frac{c}{\tan \phi}}$, this is

$$\tan \frac{\rho}{2} = c'\, e^{\frac{\theta}{\tan \phi}}.$$

On the sphere this is the curve called the loxodrome.

and for two arcs cutting at right angles

$$\frac{d\rho}{d\theta}\frac{d'\rho}{d'\theta} = -\sin^2\rho.$$

The formula for area is *

$$\int\int \sin\rho\, d\rho\, d\theta.$$

We integrate first with respect to ρ, and if the area contains the origin and each radius vector meets the curve once, and only once, our expression becomes

$$2\pi - \int_0^{2\pi} \cos\rho\, d\theta.$$

The entire rotation in going around the curve is found as on page 79, and is

$$\int_0^{2\pi} \cos\rho\, d\theta.$$

Thus the area is equal to the amount by which this rotation is less than four right angles.

For example, the area of a circle of radius ρ is $2\pi(1-\cos\rho)$, and the amount of turning in going around it is $2\pi\cos\rho$. The area of the entire plane is 2π.

7. **A modified system of coördinates :**

Writing u for $\tan x$, v for $\tan y$, r for $\tan\rho$, etc., we have

$$u^2 + v^2 = r^2.\dagger$$

The equation of a line then becomes

$$au + bv + 1 = 0,$$

and the equation of a circle

$$(1 + u^2 + v^2)c^2 = (1 + au + bv)^2.$$

---

* The unit of area being properly chosen.
† The footnote on page 80 applies here also.

III. ELLIPTIC SOLID ANALYTIC GEOMETRY

We will develop far enough to get the equation of the surface of double revolution.

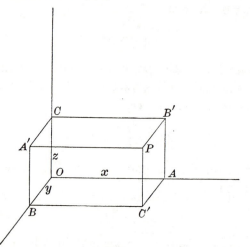

1. Coördinates, lines, and planes :

Draw three planes through the point perpendicular to the axes. For coördinates x, y, z, we take the intercepts which these planes make on the axes.

The lines of intersection of these three planes are perpendicular to the coördinate planes (Chap. I, II, 16 and 17); in fact, all the face angles in the figure are right angles except those at P and the three angles $BA'C$, $CB'A$, and $AC'B$, which are obtuse angles.

Let ρ be the radius vector to the point P, and α, β, and γ the three angles which it makes with the three axes. Then

$$\cos^2 \alpha + \cos^2 \beta + \cos^2 \gamma = 1,$$

$$\cos \alpha = \frac{\tan x}{\tan \rho}, \text{ etc.};$$

$$\tan^2 x + \tan^2 y + \tan^2 z = \tan^2 \rho.$$

For the angle between two lines intersecting at the origin

$$\cos \theta = \cos \alpha \cos \alpha' + \cos \beta \cos \beta' + \cos \gamma \cos \gamma'.$$

The angle subtended at the origin by the two points xyz and $x'y'z'$ is given by the equation

$$\cos \theta = \frac{\tan x \tan x' + \tan y \tan y' + \tan z \tan z'}{\tan \rho \tan \rho'}.$$

For the distance between two points

$$\cos \delta = \cos \rho \cos \rho' + \sin \rho \sin \rho' \cos \theta.$$

This gives us the equation of a sphere, and for $\delta = \frac{\pi}{2}$ the equation of a plane. The latter in rectangular coördinates is

$$\tan x \tan x' + \tan y \tan y' + \tan z \tan z' + 1 = 0.$$

Let p be the length of the perpendicular from the origin upon the plane, and α, β, γ the angles which this perpendicular makes with the axes. Then we have for its pole

$$\rho' = p + \frac{\pi}{2},$$

$$\tan x' = \tan \rho' \cos \alpha = -\frac{\cos \alpha}{\tan p}, \text{ etc.};$$

hence, the equation of the plane may be written

$$\cos \alpha \tan x + \cos \beta \tan y + \cos \gamma \tan z = \tan p.$$

2. The surface of double revolution :

Take one of its axes for the axis of z, suppose k the distance of the surface from this axis, and let θ denote the angle which the plane through the point P and the axis of z makes with the plane of xz. We may call z and θ latitude and longitude.

Produce OA and CB. They will meet at a distance, $\frac{\pi}{2}$, from the axis of z in a point, O', on the other axis of the surface, and there form an angle that is equal in measure to z.

From the right triangle $O'AB$

$$\cos z = \frac{\tan O'A}{\tan O'B}.$$

But $\qquad \tan O'A = \cot x,$

and $\qquad \tan O'B = \cot CB = \dfrac{\cot k}{\cos \theta}.$

Therefore, $\qquad \cos z = \dfrac{\tan k \cos \theta}{\tan x},$

or $\qquad \tan x = \dfrac{\tan k \cos \theta}{\cos z}.$

Similarly, $\qquad \tan y = \dfrac{\tan k \sin \theta}{\cos z}.$

Squaring and adding, we have for the equation of the surface

$$\tan^2 x + \tan^2 y = \tan^2 k \sec^2 z.$$

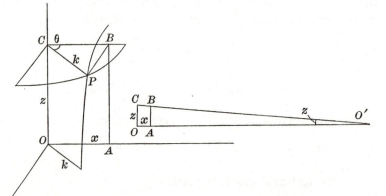

For the length of the chord joining two points on the surface, we have

$$\cos \delta = \cos \rho \cos \rho' (1 + \tan x \tan x' + \tan y \tan y' + \tan z \tan z').$$

Now, $\qquad \tan^2 \rho = \tan^2 k \sec^2 z + \tan^2 z ;$

therefore, $\qquad \sec^2 \rho = \sec^2 k \sec^2 z,$

or $\qquad \cos \rho = \cos k \cos z.$

That is, in terms of z, z', θ, and θ', we have

$$\cos \delta = \cos^2 k \cos (z' - z) + \sin^2 k \cos (\theta' - \theta).$$

From this we can get an expression for ds, the differential element of length on the surface:

$$\cos ds = \cos^2 k \cos dz + \sin^2 k \cos d\theta,$$

or, since $\qquad \cos ds = 1 - \dfrac{ds^2}{2}$, etc.,

$$ds^2 = \cos^2 k \, dz^2 + \sin^2 k \, d\theta^2.$$

z and θ are proportional to the distances measured along the two systems of circles. These circles cut at right angles, and may be used to give us a system of rectangular coördinates on the surface. The actual lengths along these two systems of circles are $\theta \sin k$ and $z \cos k$ (see Cor. p. 83). If, therefore, we write

$$\alpha = \theta \sin k, \qquad \beta = z \cos k,$$

we shall have a rectangular system on the surface where the coördinates are the distances measured along these two systems of circles which cut at right angles.

The formula now becomes

$$ds^2 = d\alpha^2 + d\beta^2.$$

An equation of the first degree in α and β represents a curve which enjoys on this surface all the properties of the straight line in the plane of the Euclidean Geometry. Through any two points one, and only one, such line can be drawn, because two sets of coördinates are just sufficient to determine the coefficients of an equation of the first degree. The shortest distance between two points on the surface is measured on such a line. For, the distance between two points on a path represented by an equation in α and β is the same as the distance between the corresponding points and on the corresponding path in a Euclidean plane in which we take α and β for rectangular coördinates. It must, therefore, be the shortest when the path is represented by an equation of the first degree in α and β. Such a line on a surface is called a *geodesic line*, or, so far as the surface is concerned, a straight line. The distance between any two points measured on one of these lines is expressed by the formula

$$d = \sqrt{(\alpha - \alpha')^2 + (\beta - \beta')^2}.$$

Triangles formed of these lines have all the properties of plane triangles in the Euclidean Geometry: the sum of the angles is π, etc. In fact this surface has the same relation to elliptic space that the boundary-surface has to hyperbolic space.

The normal form of the equation of a line is

$$\alpha \cos \omega + \beta \sin \omega = p.$$

The rectilinear generators of the surface make a constant angle, $\pm k$, with all the circles drawn around the axis which is polar to the axis of z. These generators are then "straight lines" on the surface, and their equation takes the form

$$\alpha \cos k \pm \beta \sin k = p.$$

HISTORICAL NOTE

THE history of Non-Euclidean Geometry has been so well and so often written that we will give only a brief outline.

There is one axiom of Euclid that is somewhat complicated in its expression and does not seem to be, like the rest, a simple elementary fact. It is this: *

If two lines are cut by a third, and the sum of the interior angles on the same side of the cutting line is less than two right angles, the lines will meet on that side when sufficiently produced.

Attempts were made by many mathematicians, notably by Legendre, to give a proof of this proposition; that is, to show that it is a necessary consequence of the simpler axioms preceding it. Legendre proved that the sum of the angles of a triangle can never exceed two right angles, and that if there is a single triangle in which this sum is equal to two right angles, the same is true of all triangles. This was, of course, on the supposition that a line is of infinite length. He could not, however, prove that there exists a triangle the sum of whose angles is two right angles.†

At last some mathematicians began to believe that this statement was not capable of proof, that an equally consistent

* See article on the axioms of Euclid by Paul Tannery, *Bulletin des Sciences Mathématiques*, 1884.

† See, for example, the twelfth edition of his *Éléments de Géométrie*, Livre I, Proposition XIX, and Note II. See also a statement by Klein in an article on the Non-Euclidean Geometry in the second volume of the first series of the *Bulletin des Sciences Mathématiques*.

Geometry could be built up if we suppose it not always true, and, finally, that all of the postulates of Euclid were only hypotheses which our experience had led us to accept as true, but which could be replaced by contrary statements in the development of a logical Geometry.

The beginnings of this theory have sometimes been ascribed to Gauss, but it is known now that a paper was written by Lambert,* in 1766, in which he maintains that the parallel axiom needs proof, and gives some of the characteristics of Geometries in which this axiom does not hold. Even as long ago as 1733 a book was published by an Italian, Saccheri, in which he gives a complete system of Non-Euclidean Geometry, and then saves himself and his book by asserting dogmatically that these other hypotheses are false. It is his method of treatment that has been taken as the basis of the first chapter of this book.†

Gauss was seeking to prove the axiom of parallels for many years, and he may have discovered some of the theorems which are consequences of the denial of this axiom, but he never published anything on the subject.

Lobachevsky, in Russia, and Johann Bolyai, in Hungary, first asserted and proved that the axiom of parallels is not necessarily true. They were entirely independent of each other in their work, and each is entitled to the full credit of this discovery. Their results were published about 1830.

It was a long time before these discoveries attracted much notice. Meanwhile, other lines of investigation were carried on which were afterwards to throw much light on our subject, not, indeed, as explanations, but by their striking analogies.

Thus, within a year or two of each other, in the same journal (Crelle) appeared an article by Lobachevsky giving

* See *American Mathematical Monthly*, July–August, 1895.

† The translation of Saccheri by Halsted has been appearing in the *American Mathematical Monthly*.

the results of his investigations, and a memoir by Minding on surfaces on which he found that the formulæ of Spherical Trigonometry hold if we put ia for a, etc. Yet these two papers had been published thirty years before their connection was noticed (by Beltrami).

Again, Cayley, in 1859, in the *Philosophical Transactions*, published his Sixth Memoir on Quantics, in which he developed a projective theory of measurement and showed how metrical properties can be treated as projective by considering the anharmonic relations of any figures with a certain special figure that he called the absolute. In 1872 Klein took up this theory and showed that it gave a perfect image of the Non-Euclidean Geometry.

It has also been shown that we can get our Non-Euclidean Geometries if we think of a unit of measure varying according to a certain law as it moves about in a plane or in space.

The older workers in these fields discovered only the Geometry in which the hypothesis of the acute angle is assumed. It did not occur to them to investigate the assumption that a line is of finite length. The Elliptic Geometry was left to be discovered by Riemann, who, in 1854, took up a study of the foundations of Geometry. He studied it from a very different point of view, an abstract algebraic point of view, considering not our space and geometrical figures, except by way of illustration, but a system of variables. He investigated the question, What is the nature of a function of these variables which can be called element of length or distance? and found that in the simplest cases it must be the square root of a quadratic function of the differentials of the variables whose coefficients may themselves be functions of the variables. By taking different forms of the quadratic expressions we get an infinite number of these different kinds of Geometry, but in most of them we lose the axiom that bodies may be moved about without changing their size or shape.

Two more names should be included in this sketch, — Helmholtz and Clifford.   These did much to make the subject popular by articles in scientific journals.   To Clifford we owe the theory of parallels in elliptic space, as explained on page 68. He showed that we can have in this Geometry a finite surface on which the Euclidean Geometry holds true.*

The chief lesson of Non-Euclidean Geometry is that the axioms of Geometry are only deductions from our experience, like the theories of physical science.   For the mathematician, they are hypotheses whose truth or falsity does not concern him, but only the philosopher.   He may take them in any form he pleases and on them build his Geometry, and the Geometries so obtained have their applications in other branches of mathematics.

The "axiom," so far as this word is applied to these geometrical propositions, is not "self-evident," and is not necessarily true.   If a certain statement can be proved, — that is, if it is a necessary consequence of axioms already adopted, — then it should not be called an axiom.   When two or more mutually contradictory statements are equally consistent with all the axioms that have already been accepted, then we are at liberty to take either of them, and the statement which we choose

---

* Some of the more interesting accounts of Non-Euclidean Geometry are: *Encyclopedia Britannica*, article "Measurement," by Sir Robert Ball. *Revue Générale des Sciences*, 1891, "Les Géométries Non-Euclidean," by Poincaré.   *Bulletin of the American Mathematical Society*, May and June, 1900, "Lobachevsky's *Geometry*," by Frederick S. Woods.   *Mathematische Annalen*, Bd. xlix, p. 149, 1897, and *Bulletin des Sciences Mathématiques*, 1897, "Letters of Gauss and Bolyai"; particularly interesting is one letter in which Gauss gives a formula for the area of a triangle on the hypothesis that we can draw three mutually parallel lines enclosing a finite area always the same.   The last two articles refer to the publications of Professors Engel and Stäckel, which give in German a full history of the theory of parallels and the writings and lives of Lobachevsky and Bolyai.   See also the translations by Prof. George Bruce Halsted of Lobachevsky and Bolyai and of an address by Professor Vasiliev.

becomes for our Geometry an axiom. Our Geometry is a study of the consequences of this axiom.

The assumptions which distinguish the three kinds of Geometry that we have been studying may be expressed in different forms. We may say that one or two or no parallels can be drawn through a point; or, that the sum of the angles of a triangle is equal to, less than, or greater than two right angles; or, that a straight line has two real points, one real point, or no real point at infinity; or, that in a plane we can have similar figures or we cannot have similar figures, and a straight line is of finite or infinite length, etc. But any of these forms determines the nature of the Geometry, and the others are deducible from it.

A CATALOG OF SELECTED
DOVER BOOKS
IN SCIENCE AND MATHEMATICS

Math–Geometry and Topology

ELEMENTARY CONCEPTS OF TOPOLOGY, Paul Alexandroff. Elegant, intuitive approach to topology from set-theoretic topology to Betti groups; how concepts of topology are useful in math and physics. 25 figures. 57pp. 5⅜ x 8½. 60747-X

COMBINATORIAL TOPOLOGY, P. S. Alexandrov. Clearly written, well-organized, three-part text begins by dealing with certain classic problems without using the formal techniques of homology theory and advances to the central concept, the Betti groups. Numerous detailed examples. 654pp. 5⅜ x 8½. 40179-0

EXPERIMENTS IN TOPOLOGY, Stephen Barr. Classic, lively explanation of one of the byways of mathematics. Klein bottles, Moebius strips, projective planes, map coloring, problem of the Koenigsberg bridges, much more, described with clarity and wit. 43 figures. 210pp. 5⅜ x 8½. 25933-1

CONFORMAL MAPPING ON RIEMANN SURFACES, Harvey Cohn. Lucid, insightful book presents ideal coverage of subject. 334 exercises make book perfect for self-study. 55 figures. 352pp. 5⅝ x 8¼. 64025-6

THE GEOMETRY OF RENÉ DESCARTES, René Descartes. The great work founded analytical geometry. Original French text, Descartes's own diagrams, together with definitive Smith-Latham translation. 244pp. 5⅜ x 8½. 60068-8

PRACTICAL CONIC SECTIONS: The Geometric Properties of Ellipses, Parabolas and Hyperbolas, J. W. Downs. This text shows how to create ellipses, parabolas, and hyperbolas. It also presents historical background on their ancient origins and describes the reflective properties and roles of curves in design applications. 1993 ed. 98 figures. xii+100pp. 6½ x 9¼. 42876-1

THE THIRTEEN BOOKS OF EUCLID'S ELEMENTS, translated with introduction and commentary by Thomas L. Heath. Definitive edition. Textual and linguistic notes, mathematical analysis. 2,500 years of critical commentary. Unabridged. 1,414pp. 5⅜ x 8½. Three-vol. set. Vol. I: 60088-2 Vol. II: 60089-0 Vol. III: 60090-4

GEOMETRY OF COMPLEX NUMBERS, Hans Schwerdtfeger. Illuminating, widely praised book on analytic geometry of circles, the Moebius transformation, and two-dimensional non-Euclidean geometries. 200pp. 5⅝ x 8¼. 63830-8

DIFFERENTIAL GEOMETRY, Heinrich W. Guggenheimer. Local differential geometry as an application of advanced calculus and linear algebra. Curvature, transformation groups, surfaces, more. Exercises. 62 figures. 378pp. 5⅜ x 8½. 63433-7

CURVATURE AND HOMOLOGY: Enlarged Edition, Samuel I. Goldberg. Revised edition examines topology of differentiable manifolds; curvature, homology of Riemannian manifolds; compact Lie groups; complex manifolds; curvature, homology of Kaehler manifolds. New Preface. Four new appendixes. 416pp. 5⅜ x 8½. 40207-X

Mathematics

FUNCTIONAL ANALYSIS (Second Corrected Edition), George Bachman and Lawrence Narici. Excellent treatment of subject geared toward students with background in linear algebra, advanced calculus, physics, and engineering. Text covers introduction to inner-product spaces, normed, metric spaces, and topological spaces; complete orthonormal sets, the Hahn-Banach Theorem and its consequences, and many other related subjects. 1966 ed. 544pp. 6⅛ x 9¼. 40251-7

ASYMPTOTIC EXPANSIONS OF INTEGRALS, Norman Bleistein & Richard A. Handelsman. Best introduction to important field with applications in a variety of scientific disciplines. New preface. Problems. Diagrams. Tables. Bibliography. Index. 448pp. 5⅜ x 8½. 65082-0

VECTOR AND TENSOR ANALYSIS WITH APPLICATIONS, A. I. Borisenko and I. E. Tarapov. Concise introduction. Worked-out problems, solutions, exercises. 257pp. 5⅝ x 8¼. 63833-2

THE ABSOLUTE DIFFERENTIAL CALCULUS (CALCULUS OF TENSORS), Tullio Levi-Civita. Great 20th-century mathematician's classic work on material necessary for mathematical grasp of theory of relativity. 452pp. 5⅝ x 8¼. 63401-9

AN INTRODUCTION TO ORDINARY DIFFERENTIAL EQUATIONS, Earl A. Coddington. A thorough and systematic first course in elementary differential equations for undergraduates in mathematics and science, with many exercises and problems (with answers). Index. 304pp. 5⅜ x 8½. 65942-9

FOURIER SERIES AND ORTHOGONAL FUNCTIONS, Harry F. Davis. An incisive text combining theory and practical example to introduce Fourier series, orthogonal functions and applications of the Fourier method to boundary-value problems. 570 exercises. Answers and notes. 416pp. 5⅜ x 8½. 65973-9

COMPUTABILITY AND UNSOLVABILITY, Martin Davis. Classic graduate-level introduction to theory of computability, usually referred to as theory of recurrent functions. New preface and appendix. 288pp. 5⅜ x 8½. 61471-9

ASYMPTOTIC METHODS IN ANALYSIS, N. G. de Bruijn. An inexpensive, comprehensive guide to asymptotic methods–the pioneering work that teaches by explaining worked examples in detail. Index. 224pp. 5⅜ x 8½ 64221-6

APPLIED COMPLEX VARIABLES, John W. Dettman. Step-by-step coverage of fundamentals of analytic function theory–plus lucid exposition of five important applications: Potential Theory; Ordinary Differential Equations; Fourier Transforms; Laplace Transforms; Asymptotic Expansions. 66 figures. Exercises at chapter ends. 512pp. 5⅜ x 8½. 64670-X

INTRODUCTION TO LINEAR ALGEBRA AND DIFFERENTIAL EQUATIONS, John W. Dettman. Excellent text covers complex numbers, determinants, orthonormal bases, Laplace transforms, much more. Exercises with solutions. Undergraduate level. 416pp. 5⅜ x 8½. 65191-6

CATALOG OF DOVER BOOKS

INTRODUCTORY REAL ANALYSIS, A.N. Kolmogorov, S. V. Fomin. Translated by Richard A. Silverman. Self-contained, evenly paced introduction to real and functional analysis. Some 350 problems. 403pp. 5⅜ x 8½.   61226-0

APPLIED ANALYSIS, Cornelius Lanczos. Classic work on analysis and design of finite processes for approximating solution of analytical problems. Algebraic equations, matrices, harmonic analysis, quadrature methods, more. 559pp. 5⅜ x 8½.   65656-X

AN INTRODUCTION TO ALGEBRAIC STRUCTURES, Joseph Landin. Superb self-contained text covers "abstract algebra": sets and numbers, theory of groups, theory of rings, much more. Numerous well-chosen examples, exercises. 247pp. 5⅜ x 8½.
65940-2

QUALITATIVE THEORY OF DIFFERENTIAL EQUATIONS, V. V. Nemytskii and V.V. Stepanov. Classic graduate-level text by two prominent Soviet mathematicians covers classical differential equations as well as topological dynamics and ergodic theory. Bibliographies. 523pp. 5⅜ x 8½.   65954-2

THEORY OF MATRICES, Sam Perlis. Outstanding text covering rank, nonsingularity and inverses in connection with the development of canonical matrices under the relation of equivalence, and without the intervention of determinants. Includes exercises. 237pp. 5⅜ x 8½.   66810-X

INTRODUCTION TO ANALYSIS, Maxwell Rosenlicht. Unusually clear, accessible coverage of set theory, real number system, metric spaces, continuous functions, Riemann integration, multiple integrals, more. Wide range of problems. Undergraduate level. Bibliography. 254pp. 5⅜ x 8½.   65038-3

MODERN NONLINEAR EQUATIONS, Thomas L. Saaty. Emphasizes practical solution of problems; covers seven types of equations. ". . . a welcome contribution to the existing literature. . . . "–*Math Reviews*. 490pp. 5⅜ x 8½.   64232-1

MATRICES AND LINEAR ALGEBRA, Hans Schneider and George Phillip Barker. Basic textbook covers theory of matrices and its applications to systems of linear equations and related topics such as determinants, eigenvalues, and differential equations. Numerous exercises. 432pp. 5⅜ x 8½.   66014-1

MATHEMATICS APPLIED TO CONTINUUM MECHANICS, Lee A. Segel. Analyzes models of fluid flow and solid deformation. For upper-level math, science, and engineering students. 608pp. 5⅜ x 8½.   65369-2

ELEMENTS OF REAL ANALYSIS, David A. Sprecher. Classic text covers fundamental concepts, real number system, point sets, functions of a real variable, Fourier series, much more. Over 500 exercises. 352pp. 5⅜ x 8½.   65385-4

SET THEORY AND LOGIC, Robert R. Stoll. Lucid introduction to unified theory of mathematical concepts. Set theory and logic seen as tools for conceptual understanding of real number system. 496pp. 5⅜ x 8¼.   63829-4

TENSOR CALCULUS, J.L. Synge and A. Schild. Widely used introductory text covers spaces and tensors, basic operations in Riemannian space, non-Riemannian spaces, etc. 324pp. 5⅜ x 8¼. 63612-7

ORDINARY DIFFERENTIAL EQUATIONS, Morris Tenenbaum and Harry Pollard. Exhaustive survey of ordinary differential equations for undergraduates in mathematics, engineering, science. Thorough analysis of theorems. Diagrams. Bibliography. Index. 818pp. 5⅜ x 8½. 64940-7

INTEGRAL EQUATIONS, F. G. Tricomi. Authoritative, well-written treatment of extremely useful mathematical tool with wide applications. Volterra Equations, Fredholm Equations, much more. Advanced undergraduate to graduate level. Exercises. Bibliography. 238pp. 5⅜ x 8½. 64828-1

FOURIER SERIES, Georgi P. Tolstov. Translated by Richard A. Silverman. A valuable addition to the literature on the subject, moving clearly from subject to subject and theorem to theorem. 107 problems, answers. 336pp. 5⅜ x 8½. 63317-9

INTRODUCTION TO MATHEMATICAL THINKING, Friedrich Waismann. Examinations of arithmetic, geometry, and theory of integers; rational and natural numbers; complete induction; limit and point of accumulation; remarkable curves; complex and hypercomplex numbers, more. 1959 ed. 27 figures. xii+260pp. 5⅜ x 8½. 42804-4

POPULAR LECTURES ON MATHEMATICAL LOGIC, Hao Wang. Noted logician's lucid treatment of historical developments, set theory, model theory, recursion theory and constructivism, proof theory, more. 3 appendixes. Bibliography. 1981 ed. ix+283pp. 5⅜ x 8½. 67632-3

CALCULUS OF VARIATIONS, Robert Weinstock. Basic introduction covering isoperimetric problems, theory of elasticity, quantum mechanics, electrostatics, etc. Exercises throughout. 326pp. 5⅜ x 8½. 63069-2

THE CONTINUUM: A Critical Examination of the Foundation of Analysis, Hermann Weyl. Classic of 20th-century foundational research deals with the conceptual problem posed by the continuum. 156pp. 5⅜ x 8½. 67982-9

CHALLENGING MATHEMATICAL PROBLEMS WITH ELEMENTARY SOLUTIONS, A. M. Yaglom and I. M. Yaglom. Over 170 challenging problems on probability theory, combinatorial analysis, points and lines, topology, convex polygons, many other topics. Solutions. Total of 445pp. 5⅜ x 8½. Two-vol. set. Vol. I: 65536-9 Vol. II: 65537-7

Paperbound unless otherwise indicated. Available at your book dealer, online at **www.doverpublications.com,** or by writing to Dept. GI, Dover Publications, Inc., 31 East 2nd Street, Mineola, NY 11501. For current price information or for free catalogs (please indicate field of interest), write to Dover Publications or log on to **www.doverpublications.com** and see every Dover book in print. Dover publishes more than 500 books each year on science, elementary and advanced mathematics, biology, music, art, literary history, social sciences, and other areas.